Light Scattering
by Irregularly
Shaped Particles

Light Scattering by Irregularly Shaped Particles

Edited by

Donald W. Schuerman

Space Astronomy Laboratory
State University of New York at Albany

Plenum Press • New York and London

Library of Congress Cataloging in Publication Data

International Workshop on Light Scattering by Irregularly Shaped Particles, State
  University of New York at Albany, 1979.
  Light scattering by irregular shaped particles.

  "Proceedings of the International Workshop on Light Scattering by Irregularly
Shaped Particles held at the State University of New York at Albany, New York,
June 5–7, 1979."
  1. Light – Scattering – Congresses. 2. Aerosols – Optical properties – Con-
gresses. I. Schuerman, Donald W. II. Title.
QC976.S3I57 1979   551.5'65   79-27691
  ISBN 0-306-40421-4

Proceedings of the Workshop on Light Scattering by Irregularly Shaped Particles,
held at the State University of New York at Albany, June 5–7, 1979.

© 1980 Plenum Press, New York
A Divison of Plenum Publishing Corporation
227 West 17th Street, New York, N.Y. 10011

Printed in the United States of America

Preface

This volume contains most of the invited papers presented at the International Workshop on Light Scattering by Irregularly Shaped Particles held on June 5-7, 1979, at the State University of New York at Albany (SUNYA). Over seventy participants representing many disciplines convened to define some of the ever-increasing number of resonant light-scattering problems associated with particle shape and to relate their most recent investigations in this field.

It is obvious from the two introductory papers that an investigator's primary discipline determines his/her approach to the light-scattering problem. The meteorologist, Diran Deirmendjian, advocates an empirical methodology: to model the scattering by atmospheric aerosols, using equivalent spheres as standards, in the most efficient and simplest manner that is consistent with remote sensing, *in situ*, and laboratory data. Because of the almost infinite variety of particle shapes, he questions not only the possibility but even the usefulness of the exact solution of scattering by a *totally arbitrary* particle. The astrophysicist, J. Mayo Greenberg, is primarily concerned with the information content carried by the scattered light because this radiation is the *sole* clue to understanding the nature of interstellar dust. What measurements (polarization, color dependence, etc...) should be made to best determine a given particle characteristic (size, surface roughness, refractive index, etc...)? Thus, he considers the *physics* of the scattering process to be of paramount interest. The thirty-five papers which follow the two introductory articles and which reflect linear combinations of these two approaches are representative samples of the state-of-the-art.

It will be obvious to the reader that entire questions concerning both the uniqueness and information content of scattered light are yet to be formulated, let alone answered. The Workshop consensus was that such questions could be addressed only through research emphasis on *all* of the 16 elements of the scattering intensity (or "F") matrix. Experimenters are urged to measure as many of these elements as possible in each case; at the very least, they provide self-consistency checks. Theorists are urged to

explore the information conveyed by the individual elements; they
alone (as functions of angle and wavelength) define the limits of
our search.

The efforts of many people made this workshop possible. To
all I am grateful, especially to the members of the Scientific
Organizing Committee (Petr Chýlek, SUNYA/NCAR; Robert Fenn, AFGL;
J. Mayo Greenberg, Univ. of Leiden; Ronald Pinnick, U.S. Army ASL;
John Reagan, NASA/Univ. of Arizona; and Reiner Zerull, Ruhr Univ.,
FRG) and to Dr. Jerry L. Weinberg and his colleagues and staff at the
Space Astronomy Laboratory (SUNYA) who graciously hosted the meeting.
Susan Darbyshire handled the administrative aspects of the Workshop
in her usual competent and cheerful manner, and Barbara Calkins
patiently and carefully typed the final manuscripts for publication.
Partial funding was provided by the U.S. Army Research Office under
grant number DAAG29-79-G-0016.

Don Schuerman
Workshop Chairman

Albany, New York
October, 1979

Contents

CHAPTER 1: INTRODUCTION

CHAPTER 2: USER NEEDS

CHAPTER 3: SPECIFIC PARTICLE DESCRIPTIONS

CHAPTER 6:   EXPERIMENTAL RESULTS

CHAPTER 7:   INVERSION AND INFORMATION CONTENT

SOME REMARKS ON SCIENCE, SCIENTISTS, AND THE

REMOTE SENSING OF PARTICULATES

D. Deirmendjian

The Rand Corporation

Santa Monica, Calif. 90406

Πάν μέτρον ἄριστον. Μηδέν ἄγαν.

It is flattering to be asked to deliver the opening review paper
in this workshop. Despite my efforts I was unable to dissuade the
organizing committee from putting me on the program. Besides being
a bad speaker, I dislike irregular particles and prefer to deal with
idealized homogeneous spheres, though they do not simulate exactly
even the smallest water drops (remember surface tension?). Thus,
I am afraid you are "stuck" with me and my tendency to digress into
critical remarks in areas seemingly unrelated to our main subject.
I will try to be brief in order to allow more time for the speakers
who will follow me, and who will - I am sure - be more positive, stick
more closely to the subject, and have something worthwhile to say.

Despite my upwards of 30 years residence in this country, the
longer I live here the less I seem to understand our way of life,
our *modus vivendi* and *modus operandi*, for example in crisis creation
and crisis management in all realms, in particular in the scientific/
technical one. But I don't despair, since some of my native-born
colleagues and friends don't seem to be any better off in this
respect, that is, in understanding the society we live in, and in
particular the way science and scientists operate and are managed.

The hero in Thomas Mann's (1965) last and unfinished novel, Felix
Krull, is presented as a kind of consummate "con-man". Not quite
despicable but rather human and even likable, though one feels that
there should be something false and repugnant in him. A highly in-
sightful writer like Thomas Mann, in my opinion, is telling us what
type of personality is most likely to succeed or "get ahead" in the

1

American sense in all walks of life in today's world and that of the
immediate future. (Unfortunately for us, we will never know the end
Mann was preparing for his hero). So also in Science, it seems to
me, it is sometimes difficult to draw the line between the genuine
and the "phony". Certain so-called scientists, who happen to be extremely clever or "bright", have developed their "con-manship" to
such a high degree that they are veritable virtuosos in it, and one
cannot tell where the scientist ends and the entrepreneur takes over
in order to build a sizable business empire around himself.

All this is by way of justification for the tendency on the part
of older scientists to criticize certain pieces of work they consider
unclear and misleading. One feels that there is far too little
activity of this kind around and too much of its opposite, that is,
an attitude of uncritical acceptance of anything published in scientific journals; and that a highly developed critical faculty – a
prime attribute of any worthy scientists – is becoming atrophied for
lack of exercise. To foster and maintain a healthy and vigorous
atmosphere in science, perhaps more than in other realms, together
with the "upbeat" enthusiast and builder-up, you need the harsh
critic and tearer-down. Recently I completed a survey (Deirmendjian,
1980) on the remote sensing of atmospheric particulates by scattering
techniques, under NASA (Wallops Island) sponsorship and support.
With the concurrence of the workshop chairman, I shall here present
a very brief resume of this survey. Its subject is, after all, quite
germane to that of our workshop, in that the surveyed material is
based mostly on the assumption of spherical homogeneous particle
scattering to interpret the observations. Any failures in interpretation, therefore, will partly reflect failures or weaknesses in that
assumption.

For this survey, I had to go through an enormous volume of lit-
erature, a good portion of which I found to be of doubtful relevance
and poorly presented. The period covered was roughly the decade
from 1967 to 1976 with some relevant pieces of work, published in
1977 and 1978, thrown in when appropriate. I had to confine myself
to techniques based on classical scattering theory only, excluding
Raman and fluorescent scattering, for example. Even so, the number
of actually cited references exceeded 130. For convenience I divided
my discussion into passive and active methods depending, respectively,
on whether the incident illumination is provided by natural or man-
made sources.

To save space I shall here outline only my own subjective evalu-
ations of various methods of remote monitoring of particulates based
on scattering and refer the reader to the published version
(Deirmendjian, 1980) for details. I reiterate that what follows are
my considered personal choices, arrived at without the use of any
systematic or formal evaluation procedure. Merely for convenience
and brevity in expressing these choices I used a grading system as

follows: A (excellent or very high; highly recommended); B (very good or fairly high; recommended with some qualifications); C (good; recommended with significant limitations); D (poor or low; not recommended) in the following four categories of evaluation in the order shown:

(1) Efficiency in coverage: global, geographical, vertical atmospheric, and diurnal or temporal.

(2) Information content on aerosol loading and properties.

(3) Reliability on the basis of theoretical justification (or mathematical inversion) and/or experimental checkout of the technique.

(4) Overall rating or level of recommendation for implementation.

Consider first passive techniques:

1. Solar and stellar spectral extinction method: C; B; A; B. Geographical coverage depends on the number of ground stations involved; no vertical discrimination for fixed platforms; diurnal and temporal coverage depends on cloud cover. Information content is high for total (vertical) aerosol loading but only fair for properties such as refractive index and microstructure. Reliability on both theoretical and experimental grounds is high.

2. Solar aureole method: B; B; A; B. Coverage same as in 1, except that there is good vertical discrimination if the aureole is scanned along the sun's vertical. Measurements should be confined to several narrow spectral bands within the visible and near infrared regions. Information content and reliability are high on microstructure or particle size distribution, particularly on the large particle content but fair on physical properties such as shape and refractive index.

3. A combination of 1 and 2: B; A; A; A. Among passive techniques, a system similar to the Smithsonian Astrophysical Observatory's Bouguer–Langley or "long" method of solar spectral extinction determination in conjunction with simultaneous measurements of aureole brightness and gradient at various solar elevations, is probably the most reliable and economical indicator of particulate turbidity over long periods of time. The usefulness of this combination is very high, especially if established over a strategically distributed global network of stations, not only for recording general background turbidity levels, but also for the monitoring of unusual events or anomalies such as volcanic "dust"

incursions, their geographical migrations, and temporal variations.

4. The S/H method: C; C; B; C. This method may be derived from an analysis of the ratio of the direct solar flux S and the hemispherically integrated skylight flux H (both measured within the same spectral interval) as a function of "air mass" or the consecant of the sun's zenith angle. The ratio S/H was some time ago (Deirmendjian and Sekera, 1954) found to be rather sensitive to the degree of particulate turbidity and hence a good indicator of aerosol loading. The method has not been sufficiently investigated despite its high potential for this purpose. The geographical coverage would be similar to 1 and 2 with no vertical dis- crimination and unknown information content. There is some theoretical justification, but we need more observational checks to evaluate its sensitivity to variations in the composition, amount, and microstructure of the responsible aerosols.

     There are other passive techniques, proposed or at- tempted, which I examined but did not grade. One of these involves measurements of the *diffuse reflection of sunlight by the atmosphere over water* which could be obtained aboard a geosynchronous or other space platform. Its use is attractive, but the interpretation or inversion of data is fraught with difficulties. The need to specify the sea state independently severely limits the method's usefulness for the remote sensing of turbidity. Measurements over land are excluded.

     The interpretation or inversion of *twilight sky bright- ness and polarization measurements from the earth's surface* is too uncertain for them to be of real value as a moni- toring technique, except for the detection of anomalies such as produced by volcanic eruptions. On the other hand, *sat- ellite photometry (and polarimetry) of the earth's terminator* may be a good detection and monitoring tool for mesospheric particulates such as found in noctilucent clouds.

Let us now consider active techniques:

5. Ground-based laser radar: B; C; B; B. Geographical coverage is similar to other ground based techniques. Vertical coverage is excellent if only the stratification and relative magnitude of turbidity is of interest. Information content on size distribution and particle shape may be improved by use of multiwavelength and polarimetric systems provided the necessary theoretical ground work is more fully developed than at present.

6. Satellite-borne systems, both passive and active: Such systems designed to be carried on unmanned satellites or manned space vehicles (*e.g.*, shuttles), cannot yet be fully

evaluated and they remain to be tested. There is not doubt
that space-borne systems may offer the best means for the
global monitoring of particulates, but as of now one may
estimate only their normal optical thickness and hence their
relative loading, but not their size distribution and composition - hence their true mass loading - from space observations alone.

None of the above-mentioned technqiues, by itself, is capable of
yielding complete and reliable information on the microstructure of
the aerosol population and the shape and composition of the particles.
Only some combination of passive and active techniques, ground-based
and satellite-borne, may provide the most reliable and cost effective
system for the remote monitoring of atmospheric particulates. At
present I think it would be difficult to justify an overly sophisti-
cated and expensive particulate monitoring system, in the absence of
more solid evidence than currently available on their climatological
significance. However, there is no doubt that efforts to obtain
reliable data on the ubiquitous aerosols can be justified on their
own merits regardless of climatological considerations, for example,
for verification of, and further progress in, our understanding of
the physics and chemistry of aerosols. At this stage if an operational
monitoring system were required I would be inclined to favor a judicious combination of direct solar and aureole narrow band photometry,
in combination with a ground-based lidar system and possibly a satellite-borne system, preferably employing passive techniques.

Now coming to the main subject of this workshop I think it would
be helpful if we could define a set of basic concepts and parameters
in general scattering theory and perhaps even agree on a uniform set
of symbols to represent them. Deirmendjian (1969) used a system of
basic parameters and symbols which are essentially a combination of
those adopted by van de Hulst (1957) and by Chandrasekhar (1960). A
number of them are suitable only for homogeneous spherical particles,
and I am sure some of them need to be modified or redefined for use
in the treatment of irregular particle scattering. For example, we
certainly need to agree upon a general definition for extinction
(scattering, absorption) efficiency for particles not having spherical
symmetry in all relevant properties.

In a possible follow-up workshop or topical conference on this
subject, one would like to see, among others, an examination and
discussion of the question of whether it is worth going through all
the detailed mathematical labor involved in seeking exact solutions
for scattering from non-spherical particles of an *arbitrary* degree of
irregularity and inhomogeneity.

As the Greek maxims in the epigraph are meant to convey, in
everything exaggeration and extremes should be shunned. Thus, if
the sphere assumption is perhaps too idealized (though it has certainly helped enormously in our understanding and applications of

scattering phenomena) I believe consideration of scattering by *arbi-trarily* irregular particles is the other extreme which, even if solved, cannot be of much use, for obvious reasons. As I have said elsewhere (Deirmendjian, 1969), the results of spherical particle scattering theory, besides being aesthetically highly appealing, as are - perhaps to a lesser degree - those for prolate and oblate spheroids (Asano and Yamamoto, 1975), may serve as standards against which one may define and analyze the scattering behavior - theoreti-cally predicted or experimentally observed - of various types of non-spherical particles.

This kind of comparison should be particularly useful in poly-dispersions, it seems to me, since that is the most likely type of population to be found in natural as well as in some artificial aerosols. Furthermore, polydispersity may be subdivided into various classes of mixed populations of regular (say spherical) and irregular particles in various proportions, or of particles with various chemi-cal compositions, etc. In such mixed populations parameters defining departures from sphericity could be usefully defined. Experimental measurements of all relevant scattering parameters are, of course, of great importance here, and the more independent sets of measurements available, the better it would be for our understanding and advance-ment of the field.

At any rate there is no doubt that the study of the scattering properties of irregular particles is here to stay, whether one likes them or not, since they certainly exist, both in planetary atmosphere and in circumsolar and interstellar space. A workshop like the one we are attending is, therefore, an appropriate recognition of this fact.

REFERENCES

Asano, S. and Yamamoto, G., 1975, *Appl. Opt.*, 14, 29.

Chandrasekhar, S., 1950, "Radiative Transfer," Clarendon, Oxford, p. 393; Dover Edition, 1960.

Deirmendjian, D., 1980, A Survey of Light Scattering Techniques Used in the Remote Monitoring of Atmospheric Aerosols, (accepted for publication in *Rev. Geophys. Sp. Phys.*).

Deirmendjian, D., 1969, "Electromagnetic Scattering on Spherical Polydispersions," American Elsevier, N. Y., p. 290.

Deirmendjian, D. and Sekera, Z., 1954, *Tellus*, 6, 382.

Mann, Thomas, 1965, "Confessions of Felix Krull, Confidence Man; The Early Years," Translated from the German by Denver Lindsley, Modern Library, N. Y., p. 378.

van de Hulst, H. C., 1957, "Light Scattering by Small Particles," Wiley, N. Y., p. 470.

FOCUSING IN ON PARTICLE SHAPE

J. Mayo Greenberg

Leiden University, Laboratory Astrophysics

Leiden, Netherlands

1. INTRODUCTION

While it is recognized that naturally occurring scattering particles are not always perfect spheres, nevertheless a vast amount of theoretical calculations have been made for perfect spheres. Why is this? There are two obvious and very important justifications: (1) many of the qualitative and even quantitative characteristics of particle scattering are adequately represented by equivalent spheres – where equivalence may have varying definitions in applications to different scattering properties, and (2) Mie theory calculations have been easy to perform even before the advent of modern high speed computers – dating back to the 1930's.

However, an uncritical use of Mie theory is known to all of us here to lead, in a number of applications of practical interest, to incomplete and in some cases to spurious and misleading results. I do not just mean such obvious examples as the explanation of polarization of radiation passing through aligned elongated particles. There are many other less obvious cases such as the ratio of forward to back scattering depending on whether or not the particles are spheres. Just what these problems are and how we might best tackle them theoretically and experimentally is the task of all you people here. I will merely try in these introductory remarks to summarize what appear to me to be some of the principal guiding themes with the expectation that during the course of this workshop the definitions of the problem areas will not only have been sharpened but also extended.

2.   EXAMPLES OF NON-SPHERICAL PARTICULATES

The following is a representative selection of physical situations in which there is need for scattering results which cannot
simply be inferred from Mie theory calculations:
a)   Atmospheric aerosols and smokes.—There will be numerous
detailed statements on this area at this meeting. Needless to say,
the particle types are vast - from simple smooth nonsphericity to
agglomerates as *seen* in photographs.
b)   Dust in planetary atmospheres.
c)   Interstellar dust.—The fact that the light of distant stars
is polarized as a result of differential extinction for different
alignments of the electric vector of the radiation leads us to infer
definitively that the scattering particles *have* to be non-spherical.
d)   Interplanetary dust.—Measurements of the size of interplanetary particles and observations of the zodiacal light - the
scattered sunlight off of these particles - leads to the necessity
that these particles *not* be regular in shape but must have inhomo-
geneities at least in their surface. In addition, particles which
are collected in the upper atmosphere (and which appear to be of
nonterrestrial origin) always show aggregated characteristics.
e)   Comet dust.—It is clearly unreasonable to expect that the
dust which is the debris of a comet as it sheds material in its
approach to the sun can be spherical. Furthermore, the theories of
comet formation lead us to expect that the basic comet material is
aggregated interstellar dust and as such is certainly inhomogeneous.
f)   Macromolecules.

3.   PHYSICAL CHARACTERIZATION OF PARTICULATES

In Figure 1 there is a schematic drawing of an arbitrary non-
spherical but homogeneous particle. The key parameters of the

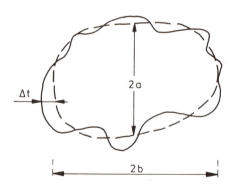

FIG. 1.   Key parameters used to describe an arbitrary non-spherical
but homogeneous particle.

particle.which govern its scattering characteristics are described by:

Size - x = 2πa/λ: a is a characteristic length of the particle and λ is the wavelength of the radiation.

Shape - elongation: b/a where b and a are representative of the length and width. Appropriate definition for flatness.

       roughness: $\Delta x_{max}/x$ or $\overline{\Delta x}/x$ where $\Delta x = 2\pi\Delta t/\lambda$ and Δt is the variation from a smoothed particle boundary (defined in some appropriate way). $\overline{\Delta x}$ is the mean deviation from smoothness defined as $\overline{\Delta x} = \{<(\Delta x)^2>\}^{1/2}$.

       edges: Sharpness of surface deviation.

Material - index of
      refraction: $m = m' - im''$ (or $n - ik$) where m' is the real (refractive) part of the index and m'' is the imaginary (absorptive) part of the index.

      dielectric: Either clear or slightly absorbing in the visual with absorption edges in the ultraviolet and broad absorption lines in the infrared. m' approximately constant in the visual, m'' << 1 in the visual.

      metallic: Absorbing from the infrared to the ultraviolet. m' and m'' $\gtrsim 1$ (very large in the infrared).

As difficult as it is to give a general and simple parametric representation of homogeneous particles, the situation is almost impossible for inhomogeneous particles with some few exceptions. The following examples can be extended as needed:

Shell structure: Some particles consist of a core of one material surrounded by a mantle of some other material. This may be true for aerosols as well as for interstellar particles. The relevant parameters here are the relative thicknesses of core and mantle.

"Raisin pudding": Particles may grow by aggregation of some basic matrix material in which are collected (or imbedded) inclusions which are not close to each other. Here the relevant parameter is the size and distance of the inclusions.

Aggregates: As distinguished from "raisin pudding" particles, aggregating material consists itself of particles so that the distance between the aggregated particles is comparable with the particle sizes.

Birefringence: Different index of refraction for different directions of polarization. Optical activity: rotation of plane of polarization.

4.   SCATTERING PARAMETERS

The primary effects of a particle on the impinging light are indicated in Figure 2. Radiation is either scattered and/or absorbed. The former may be directly observed; the latter is usually inferred unless the absorption is spectroscopic. We summarize below the basic scattering terms. See any standard textbook like those of Kerker (1969) or van de Hulst (1957) for details.

Scattering amplitude: $\mathbf{A}(\theta,\phi) = \mathbf{A}_1(\theta,\phi) + \mathbf{A}_2(\theta,\phi)$ is the vector amplitude of the scattered radiation with the component indexes (1 and 2) denoting the directions perpendicular and parallel to the scattering plane (see Fig. 2).

Angular scattering intensity: $|\mathbf{A}(\theta,\phi)|^2 = |\mathbf{A}_1|^2 + |\mathbf{A}_2|^2 = i_1(\theta,\phi) + i_2(\theta,\phi)$.

Polarization: $P(\theta,\phi) = \dfrac{i_1(\theta,\phi) - i_2(\theta,\phi)}{i_1(\theta,\phi) + i_2(\theta,\phi)}$.

Backscatter: $i(\pi)$.

Total scattering cross section: $C_{sc} = k^{-2} \int\!\!\int |\mathbf{A}(\theta,\phi)|^2 \sin\theta d\theta d\phi$.

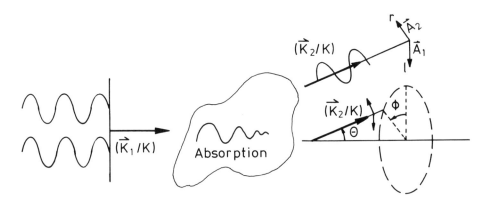

FIG. 2.   A schematic representation of the scattering process. Radiation in the incident direction ($\mathbf{k}_1/k$) is either absorbed and/or scattered in a direction ($\mathbf{k}_2/k$) parameterized by the scattering angle θ and the azimuthal angle ϕ. The vector amplitudes $\mathbf{A}_1$ and $\mathbf{A}_2$ are parallel and perpendicular, respectively, to the scattering plane defined by $\mathbf{k}_1$ and $\mathbf{k}_2$.

Scattering asymmetry: $g = \iint \dfrac{C_{sc}(\theta,\phi)}{C_{sc}} \cos\theta \, \sin\theta d\theta d\phi = \langle\cos\theta\rangle$.

Absorption cross section: C_{abs}.

Total (extinction) cross section: $C_t = C_{abs} + C_{sc}$.

Albedo: $\mathbf{a} = C_{sc}/C_t$.

Radiation pressure (for spherical particles): $C_{pr} = C_{abs} + (1-g)\, C_{sc}$.

5.   SPHERE VERSUS NON-SPHERE

The problem is to measure the basic scattering parameters in section 4 and to map these on the basic particle parameters in section 3. This is the inversion problem. Knowing how difficult it is to map 4 on 3 for spheres alone; *i.e.*, eliminating the varieties of shape and inhomogeneity entirely, we see immediately that our task is enormous. However, if we thought it was hopeless we wouldn't be here. There are two ways of getting into the problem; one is to have auxiliary knowledge from other sources which allows the pre-assigning of some particle characteristics – shape, material – and then deal with the remainder; the other is to try to separate the effects of varying one type of parameter at a time while keeping the others constant. A typical example of the latter is the assembling of a library of sphere scattering characteristics and then attempting to quantify the effects of smooth deviations from the spherical shape in order to apply the Mie theory results to arbitrary situations. This has proved in some cases to be very successful, but it has some pitfalls. I need mention only two to indicate both an application and a difficulty of this method.

It turns out that, given the back scattering from remote sensing of a distribution of particles, we would underestimate the total scattering cross section of the distribution if we had assumed them to be spheres when, in fact, they were spheroids or any other non-spherical shape. This is because the radar cross sections for spheres are, in the mean, larger than for *any* slightly non-spherical particle while a modest deviation from nonsphericity produces only small effects on the angular distribution at all other angles and, consequently, on the total cross section.

On the other hand, we are now finding that the zodiacal light may, on the basis of the limited scattering information, be explained not only by the classically assumed spherical particles in the micron size range but also by at least two distinctly different varieties of highly irregular particles in the 10-100 micron size

range! In this field, a great deal more observational as well as
theoretical information will be required before the problems have
even a chance of being solved uniquely.

To sum up, it is highly unlikely or impossible that a purely
analytical approach can be used to solve the inversion problem and
we will have to remain satisfied with at least a partially synthetic
approach. This means that we must continue to accumulate a broad
organized body of knowledge of the scattering by particles of various
shapes, sizes, etc.

6. THE SCATTERING BY INDIVIDUAL PARTICLES

We shall hear a number of talks on both theoretical and experimental methods of solving the scattering problem for individual
particles. Therefore, I will not attempt to be complete in my survey.
I will, however, try to point out the main lines of attack which are
being pursued.

Exact theoretical solutions are available for the homogeneous
sphere, layered sphere, homogeneous circular cylinder, layered circular cylinder. Ellipsoidal cylinders may also be solved by separation of variables. Although exact solutions are not available in
a classical sense for the homogeneous spheroid, the solutions may be
put in a form which is numerically manageable by high speed digital
computers to a high order of accuracy over an ever increasing size
range.

Experimental solutions are available for problems which, because of gross irregularity or inhomogeneity, are not amenable to
computer techniques. The microwave analog method has been developed
at Albany and at Bochum to solve the scattering by a wide range of
particle shapes, inhomogeneities, and indexes of refraction. There
are also being developed particle suspension techniques which may
make it possible to study directly the scattering by naturally
occurring particles.

All of the above methods have advantages as well as limitations.
Given an arbitrary particle, the exact methods are applicable only
if equivalent spheres, spheroids, or cylinders may reasonably represent their basic scattering characteristic as *observed*. The microwave analog methods are limited in their simulation of indexes of
refraction, particularly in cases involving spectral absorption
characteristics. They are also limited in the maximum size to wavelength which may be simulated. The Microwave Laboratory at Albany
is, however, capable of providing *all* scattering properties of those
particles it *can* simulate. By "all" we mean not only the angular
scattering distribution but also the total cross section and, by
subtracting the integral of the former from the latter, the

absorption cross section. This means that we can obtain the *albedo* for particles for which no theoretical methods yet exist. The albedo is a very important parameter in radiative transfer calculations such as are needed to understand the temperature balance in the Earth's and other planetary atmospheres.

Finally, there are the approximate theoretical methods. These methods usually take advantage of a limitation in value or size or some parameter of the scatterer. For example, $x \ll 1$ (or more accurately $mx \ll 1$) leads to the so-called Rayleigh approximation; $m'-1 \ll 1$, $m'' \ll 1$ leads to the eikonal (or WKB) type of approximation; $\Delta x \ll 1$ (on a sphere) leads to perturbation methods.

In all solutions and in interpretations of observations it is of great help to bear in mind the fundamental scattering relationships, particularly the symmetry relationships. The following items have provided the basis for many useful qualitative as well as quanitative ways of understanding the scattering problems.

Starting first with the scalar wave integral formulation, for simplicity, we then give some results for the electromagnetic case. Consider an incident plane wave $\psi_{inc} = e^{-i\,\mathbf{k}_1\cdot\mathbf{r}}$ on a particle whose index of refraction is given quite generally by $m^2(\mathbf{r}) = k^2(\mathbf{r})/k^2$ where k is the incident wave number ($= 2\pi/\lambda$) and $k^2(\mathbf{r})$ is the wave number at a point $\mathbf{r}$ within the scatterer. The solution is given by

$$\psi(\mathbf{r}) = \psi_{inc} - \frac{1}{4\pi}\int \frac{e^{-ik|\mathbf{r}-\mathbf{r}'|}}{|\mathbf{r}-\mathbf{r}'|}\{k^2 - k^2(\mathbf{r}')\}\,\psi(\mathbf{r}')d^3\mathbf{r}'$$

$$(1)$$

$$\xrightarrow[r\to\infty]{} e^{-i\,\mathbf{k}_1\cdot\mathbf{r}} - \frac{1}{4\pi}\frac{e^{-ikr}}{r}\int e^{i\,\mathbf{k}_2\cdot\mathbf{r}'}\{k^2 - k^2(\mathbf{r}')\}\,\psi(\mathbf{r}')d^3\mathbf{r}',$$

from which the scattered wave in the direction $\mathbf{k}_2$ is obtained as

$$\psi_{scatt} = \frac{S(\mathbf{k}_1,\mathbf{k}_2)}{ikr}\,e^{-ikr}\quad,\text{ and}$$

$$(2)$$

$$S(\theta,\phi) \equiv S(\mathbf{k}_1,\mathbf{k}_2) = -\frac{ik}{4\pi}\int e^{i\mathbf{k}_2\cdot\mathbf{r}'}\{k^2 - k^2(\mathbf{r}')\}\,\psi(\mathbf{r}')d^3\mathbf{r}' \ .$$

The physical interpretation of this is that the scattered wave is made up of a sum of waves radiating from different points within the scatterer with strengths proportional to a product of the amplitude of the penetrating wave and the difference of the scatterer from free space. This physical picture provides the basis for such approximations as the R.G.B., the eikonal and even the Rayleigh approximations.

From equation (1) one may derive the optical theorem from conservation of energy (or particles) to be

$$- (4\pi/k^2)\mathrm{Re}\{S(\mathbf{k}_1,\mathbf{k}_2)\} = k^{-2}\int |S(\mathbf{k}_1,\mathbf{k}_2)|^2 \, d\Omega_{\mathbf{k}_2}$$

$$+ k^{-1}\int |\psi|^2 \, \mathrm{Im}\, \{k^2 - k^2(\mathbf{r})\} \, d^3\mathbf{r} \quad . \tag{3}$$

The first term on the r.h.s. of equation (3) is the scattering cross section, and the second term is the absorption cross section so that what we have may be summarized by

$$- (4\pi/k^2)\mathrm{Re}\{S(\mathbf{k}_1,\mathbf{k}_2)\} = C_t = C_{sc} + C_{abs} \quad . \tag{4}$$

The extension to the electromagnetic case is relatively straight-forward and gives for an incident electromagnetic wave

$$\mathbf{E}_{inc}(\mathbf{r}) = E\, \mathbf{q}\, e^{-i\mathbf{k}_1 \cdot \mathbf{r}} \quad ,$$

where $\mathbf{q}$ is a unit vector along the direction of polarization, and E is the amplitude,

$$\mathbf{E}(\mathbf{r}) \xrightarrow[|\mathbf{r}| \to \infty]{} \mathbf{E}_{inc}(\mathbf{r}) + \frac{Ee^{-ikr}}{ikr}\, \mathbf{A}(\mathbf{k}_1,\mathbf{k}_2) \quad . \tag{5}$$

The vector scattering amplitude is then given by

$$\mathbf{A}(\mathbf{k}_1,\mathbf{k}_2) = \frac{1}{4\pi}\left[\mathbf{D}(\mathbf{k}_1,\mathbf{k}_2) - \frac{\mathbf{k}_2}{k^2}\{\mathbf{k}_2 \cdot \mathbf{D}(\mathbf{k}_1,\mathbf{k}_2)\}\right], \tag{6}$$

where

$$\mathbf{D}(\mathbf{k}_1,\mathbf{k}_2) = -\frac{ik}{4\pi E}\int e^{i\mathbf{k}_2 \cdot \mathbf{r}'}\{k^2 - k^2(\mathbf{r}')\}\, (\mathbf{r}')d^3\mathbf{r}' \quad . \tag{7}$$

The vector optical theorem is then

$$C_t' = \frac{4\pi}{k^2} \, \text{Re}\{\mathbf{q} \cdot \mathbf{A_q}(\mathbf{k}_1, \mathbf{k}_2)\} \ , \tag{8}$$

and the absorption is

$$C_{abs} = \frac{1}{kE_{inc}^2} \int |\mathbf{E}|^2 \, \text{Im}\{k^2 - k^2(\mathbf{r}')\} d^3\mathbf{r}'. \tag{9}$$

In the R.G.B. approximation one assumes that the scatterer is weak so that one may approximate the internal wave by the incident wave. In this approximation the optical theorem cannot be used because, for example, for a non-absorbing scatterer one would get the ridiculous result that the total cross section is zero since the scattering amplitude is imaginary and has no real part. *However,* the sum of the scattering cross section (obtained from integrating an approximate $|\quad|^2$) and an approximate absorption cross section obtained from equation (9) would be reasonable even if one uses a trial interior wave function equal to the undisturbed initial wave (R.G.B.).

The eikonal approximation results from replacing the internal wave by one which has the same wave number as the incident wave but is phase shifted along linear undeviated paths within the scatterer. This approximation has proved to be very powerful in arriving at semi-quantitative predictions of the effects of (smooth) nonsphericity relative to spheres. Its usefulness is also enhanced by its amazing simplicity and the ease of using it to obtain a good range of analytical results.

The Rayleigh approximation is so well-known and its power so great that I need hardly mention it here. However, I felt it was interesting to present some results of its application which are not apparently well-known. These are presented in three appendices.

7. CONCLUDING REMARKS

As you can see, I have not tried to present a complete summary of what we know or don't know about non-spherical particles. This is because the titles and abstracts in the program lead me to think that we can look forward in the next few days to hear a number of interesting and new developments in the problem of non-spherical particle scattering as well as a reasonable coverage of the topics. I have not considered.

APPENDIX A: ABSORPTION BY METALLIC NEEDLES   (ELONGATED SPHEROIDS)

It is well known that the long wavelength emissivity of substantially elongated (see appendix B for the near-spherical case)
particles is greater per unit volume than that for spheres.   This
fact is occasionally assumed to imply unjustifiably remarkable properties of needles.   It is the purpose of this note to show that the
classical Rayleigh (quasi-static) approximation, which applies in
the cases of interest, defines restrictive conditions on the absorption and emission by slender particles.

The justification for presenting in some detail what is clearly
a rather elementary derivation lies in the clarification of the conclusions and their consequences.   We shall limit the derivation to
the prolate spheroid (needle) since the oblate spheroidal results
are essentially the same.   Numerical results will be given for particles whose optical properties are like those of a metal in the far
infrared and microwave region.

The absorption cross section for a prolate spheroidal particle
whose size is much smaller than the wavelength and for which the
symmetry axis is parallel or perpendicular to the polarization of
the radiation is given by

$$C_{\parallel,\perp}^{p} = -4\pi k \, \text{Im}\{\alpha_{\parallel,\perp}\} \, , \tag{A1}$$

where $k = 2\pi/\lambda$, λ = wavelength (hereafter given in microns) and the
polarizabilities are

$$\alpha_{\parallel,\perp} = \frac{V(m^2-1)}{4\pi} \{(m^2-1)L_{\parallel,\perp} + 1\}^{-1} \, , \tag{A2}$$

where V = particle volume, m^2 = complex dielectric constant = $\varepsilon_1 - i\varepsilon_2 =$
$(m'^2 - m''^2) - i2m'm''$; m' and m'' are the real and imaginary parts of
the index of refraction.

We shall consider the materials for which, as λ increases, $\varepsilon_1 =$
constant and $\varepsilon_2 = \beta\lambda$ where β is a constant ($m^2 = \varepsilon_1 - i4\pi\sigma/\omega$ for metals,
σ = conductivity).   The shape parameters are given by

$$L_{\parallel} = \frac{(1-e^2)}{e^2}\left\{-1+\frac{1}{2e}\ln\left(\frac{1+e}{1-e}\right)\right\} \, ; \quad 2L_{\perp} = 1-L_{\parallel} \, , \tag{A3}$$

where $e = \{1-(b/a)^2\}^{1/2}$ = eccentricity.   The spheroidal absorption cross sections per unit volume are

$$\frac{C_{||,\perp}^{p}}{kV} = \frac{\varepsilon_2}{1 + 2(\varepsilon_1-1)L_{||,\perp} + L_{||,\perp}^2 \{(\varepsilon_1-1)^2 + \varepsilon_2^2\}} \qquad (A4)$$

which, for the sphere (e = 0), reduces to

$$\frac{C^{s}}{kV} = \frac{9\varepsilon_2}{(\varepsilon_1+2)^2 + \varepsilon_2^2} \ . \qquad (A5)$$

We note that $(\varepsilon_1-1)^2 + \varepsilon_2^2 = |m^2-1|^2$.

For random orientation, the spheroid cross section is given by $\overline{C}^p = \{C_{||}^p + 2C_{\perp}^p\}/3$. The limiting expressions of absorption cross section for needles may be readily deduced. As $\varepsilon \to 1$, $(b/a) \to o$ and the shape parameters $L_{||}$ and $L_{\perp}$ approach

$$L_{||} \equiv \left(\frac{b}{a}\right)^2 f \simeq \frac{b^2}{2a^2}\left\{-\ln 2 + \ln\left(\frac{a}{b}\right)\right\} \xrightarrow[\varepsilon \to 1]{} 0$$

$$L_{\perp} \xrightarrow[\varepsilon \to 1]{} \simeq \frac{1}{2}$$

where f is of the order of unity in the cases of interest.

At first glance there may not appear to be an essential difference between the spheroid (eq. A4) and sphere (eq. A5) emissivities as $\lambda \to \infty$ ($\varepsilon_2 \to \infty$). If, however, for a given value of λ we require the spheroid to be slender enough to satisfy

$$L_{||}^2|m^2-1|^2 \lesssim 1 \ , \qquad (A7)$$

we then see how it is possible for the needle-like spheroids to have much higher emissivities than spheres for long wavelength (Greenberg 1972).

As $\lambda \to \infty$, $|m^2-1|^2 \to 4m'^2$, and $m' \sim (\varepsilon_2/2)^{1/2}$. Using this, the long wavelength limit expression for the condition given in equation (A7) is

$$L_{||}^2\beta^2\lambda^2 \lesssim 1 \ . \qquad (A7')$$

This establishes a lower bound on the elongation of the spheroid for which large cross sections may be expected.

An upper bound on the length - the elongation for a given thickness of the spheroid - is determined by the condition for the Rayleigh approximation in the *length*. The reason for applying this criterion for limiting the particle size is that, when it is not satisfied, the cross sections given by equation (A4) are overestimates of the true cross sections. It is only when *all* elements within the particle radiate in phase with each other that we can achieve the high emissivities. The condition stating this small phase difference along the length of the particle is

$$m'2\pi a/\lambda \ll 1 \ (\simeq 0.1) \qquad\qquad (A8)$$

as inferred from the expression for spheres (see, *e.g.*, van de Hulst, 1957). Combining equations (A6), (A7') and (A8), the bounds on the particle length, 2a, for a given thickness, 2b, are expressed by the inequality

$$f^{1/2}\beta^{1/2}b\lambda^{1/2} \lesssim a \lesssim 0.022\beta^{-1/2}\lambda^{1/2} \qquad\qquad (A9)$$

where we have simplified by going to the long wavelength limit, $m' \to m'' \simeq (\varepsilon_2/2)^{1/2}$.

The inequality (A9) is meaningful only when the ratio of the upper to the lower bound is greater than unity, that is

$$f^{1/2} < 0.022 \ \beta^{-1}b^{-1} \ , \qquad\qquad (A10)$$

which is valid for all (long) wavelengths and therefore applies equally well to the emission of cold dust and of dust in H II regions. Equation (A10) finally produces the conditions on particle elongation as

$$(a/b) \lesssim 2 \ \exp\{ \ 2(0.022 \ \beta^{-1}b^{-1})^2\} = 2 \ \exp\{\beta^{-2}b^{-2}10^{-3}\}. \qquad (A11)$$

As a numerical example, let us represent the complex index of refraction by $\varepsilon = 4$, $\beta = 10$ which roughly corresponds to graphite in the far infrared (Greenberg, 1968). As we see in Table A1, for $b = 0.01$ μm the maximum particle elongation is only 2 which is not large enough to produce significantly larger emissivities than for

spheres. We must consider particles with semi-thickness in the range
b = 0.001 μm = 10 Å before any interesting consequences may appear.
For the latter size the maximum particle elongation is $a/b = 3 \times 10^4$.

The maximum wavelength corresponding to the maximum elongation
is obtained from equation (A7')

$$\lambda \leqslant L_{\|}^{-1} \beta^{-1} \tag{A12}$$

Judging by the results given in the Table, there does not appear to
be any obvious way to get extremely high emissivities at millimeter
wavelengths except for conducting particles as small or smaller than
b = 10Å. It is no longer clear that for this size classical electromagnetic scattering applies so that the entire theoretical basis
should be reconsidered quantum mechanically. Finally, does a particle which is 20 Å thick and 60 μm long have sufficient rigidity to
exist if it is subject to collisional (rotational) effects in space?

TABLE A1

MAXIMUM RATIO OF SPHEROID TO SPHERE ABSORPTION PER UNIT VOLUME

Elongation = a/b; dielectric coefficient $\varepsilon = 4 - 10i\lambda$.

b (μm)	(a/b)max	λmax (μm)	L	$\dfrac{c^p/V}{c^s/V}$
0.01	2.2	0.9	0.156	30
0.003	5.9	2.3	4.3×10^{-2}	400
0.001	3.2×10^4	2.1×10^7	4.7×10^{-9}	4×10^{14}

APPENDIX B: DOES THE SPHERE GIVE THE LEAST ABSORPTION PER UNIT VOLUME FOR SMALL DEVIATIONS FROM SPHERICITY?

Consider a prolate spheroid as an example. We may, in the limit of small deviation from the sphere, write the shape parameters $L_{||}$ and $L_{\perp}$ (as defined in Appendix A) as

$$L_{||} = 1/3 + \ell_p$$
$$L_{\perp} = 1/3 - \ell_p/2 \;,$$

(B1)

where $\ell_p \to 0$ for the sphere as the ellipticity, $e = (1-b^2/a^2)^{1/2}$, approaches 0. The quantity ℓ_p may be expanded in a series in e as

$$\ell_p = \sum_{n=1}^{\infty} \frac{2(n+2)}{(2n+1)(2n+3)} e^{2n} \;.$$

(B2)

The cross section per unit volume for the sphere is a maximum or minimum if, at the same time,

$$\frac{\partial (C_{||,\perp}^P/V)}{\partial e}\Big|_{e=0} = \frac{\partial (C_{||,\perp}^P/V)}{\partial \ell_p} \frac{d\ell_p}{de}\Big|_{e=0} = 0 \;,$$

(B3a)

and

$$\frac{\partial^2 (C_{||,\perp}^P/V)}{\partial e^2}\Big|_{e=0} \lessgtr 0 \left\{ \begin{array}{l} \text{maximum} \\ \text{minimum} \end{array} \right. .$$

(B3b)

Expanding $\alpha_{||}$ and $\alpha_{\perp}$ (see equation A2) in a power series in ℓ_p we obtain

$$\frac{4\pi\alpha_{||}^P}{V(m^2-1)} = \left[\frac{1}{3}(m^2+2)^{-1}\right]\left\{1 - \frac{3(m^2-1)}{(m^2+2)}\ell_p + \left[\frac{3(m^2-1)}{(m^2+2)}\right]^2\ell_p^2 + \ldots\right\}$$

(B4)

$$\frac{4\pi\alpha_{\perp}^P}{V(m^2-1)} = \left[\frac{1}{3}(m^2+2)^{-1}\right]\left\{1 + \frac{3(m^2-1)}{2(m^2+2)}\ell_p + \left[\frac{3(m^2-1)}{2(m^2+2)}\right]^2\ell_p^2 + \ldots\right\}.$$

Sine ℓ_p is an even function of e, we note that $(\partial\ell_p/\partial e)_{e=0} = 0$, and thus equation (B3a) is automatically satisfied while equation

(B3b) becomes

$$\frac{\partial^2 (c_{\parallel,\perp}^p /V)}{\partial e^2}\Bigg|_{e=0} \approx \frac{\partial (c_{\parallel,\perp}^p /V)}{\partial \ell_p} \frac{d^2 \ell_p}{de^2}\Bigg|_{e=0} \lessgtr 0 \quad ,$$

and, since $d^2\ell_p/de^2$ is greater than zero, equation (B3b) reduces to a consideration of

$$\frac{\partial (c_{\parallel,\perp}^p /V)}{\partial \ell_p}\Bigg|_{e=0} \lessgtr 0 \left\{ \begin{array}{l}\text{maximum}\\\text{minimum}\end{array}\right. . \tag{B5}$$

Inserting the expansions for $\alpha_\parallel$ and $\alpha_\perp$ into equation (B5), we see that the first non-zero terms give

$$\frac{\partial}{\partial \ell_p}\left(\frac{C}{4\pi k V}\right)\Bigg|_{e=0} = \text{Im} \left(\frac{m^2-1}{m^2+2}\right)^2 \tag{B6a}$$

$$\frac{\partial}{\partial \ell_p}\left(\frac{C}{4\pi k V}\right)\Bigg|_{e=0} = -\frac{1}{2} \text{Im}\left(\frac{m^2-1}{m^2+2}\right)^2 . \tag{B6b}$$

After some algebra we find

$$\text{Im}\left(\frac{m^2-1}{m^2+2}\right)^2 = -\frac{6\varepsilon_2}{[(\varepsilon_1+2)^2-\varepsilon_2^2]^2} [\varepsilon_1^2+\varepsilon_1-2+\varepsilon_2^2] \quad , \tag{B7}$$

so that $(C_\parallel/V)$ is a maximum or a minimum depending on whether

$$\left(\varepsilon_1^2+\varepsilon_1-2+\varepsilon_2^2\right) \begin{array}{l}> 0 \quad \text{max}\\< 0 \quad \text{min}\end{array} \quad , \tag{B8}$$

and *vice versa* for $(C_\perp/V)$.

Thus, for perfectly aligned spheroids, equation (B5) implies *either* a maximum or a minimum depending on both *the alignment* and the *index of refraction*.

On the other hand, we note that for randomly aligned spheroids in either polarized or unpolarized radiation we may write $<C> = (C_{||}+2C_\perp)/3$ so that the coefficient of ℓ_p in $<C>$ (see equation B6) is identically zero leaving only even powers of ℓ_p in the expansion of $<C>$. This implies that, for this case, the sphere represents a point of inflection.

We may rewrite equation (B8) in the form

$$
\left((\varepsilon_1+ \tfrac{1}{2})^2 + \varepsilon_2^2\right)
\begin{matrix} > (3/2)^2 \\ < (3/2)^2 \end{matrix}
\quad
\begin{matrix} C_{||}/V \\ \hline \text{max} \\ \text{min} \end{matrix}
\quad
\begin{matrix} C_\perp/V \\ \hline \text{min} \\ \text{max} \end{matrix}
\qquad (B9)
$$

so that all we need look for is whether the pair of values ε_1, ε_2 lie inside or outside the circle of radius 3/2 displaced - 1/2 units along the ε_1 axis. Of course, only $\varepsilon_2 > 0$ is allowed from conservation of energy; *i.e.* real absorption.

APPENDIX C: RADIATION PRESSURE ON NON-SPHERICAL PARTICLES

It is usually believed that when a particle is subjected to radiation from a particular direction it will be subject to a force along that same direction. This happens to be identically true only for spheres and for particles with axial symmetry about the direction of propagation. As a matter of fact, it is obvious, upon a little thought, that if the radiation is not scattered symmetrically relative to a plane containing the propagation direction, there will be a net or resultant radiation force normal to this plane.

There are two cases for which explicit forces perpendicular to the direction of propagation may be readily calculated. These are the small spheroid (Rayleigh approximation) and the infinite circular cylinder.

Let us consider the prolate spheroid as an example. Suppose the spheroid axis is at an angle ψ with respect to the incident direction. Then the induced dipole moment component for radiation polarized within the plane containing the particle axis are

$$p_{||} = E\sin\psi \; \alpha_{||}$$

$$p_\perp = E\cos\psi \; \alpha_\perp$$

along and perpendicular to the particle axis.

Since $\alpha_{||} \neq \alpha_{\perp}$ (and generally, but not always, $\alpha_{||} > \alpha_{\perp}$), the total dipole moment is not perpendicular to the radiation direction. The induced oscillating dipole radiates symmetrically about itself and therefore *not* about the perpendicular to the radiation direction. In the limit of an infinitely slender spheroid $\alpha_{\perp}/\alpha_{||} \rightarrow 0$, and therefore, the radiation force for zero absorption is exactly perpendicular to the particle axis. This limit applies also to the infinite circular cylinder.

BIBLIOGRAPHY

The following is obviously not a complete list of references. It is intended as a source of background information. A few papers are included in which additional references to important works may be found. Since a good deal of my own learning has come from working with students, I have listed their Ph. D. dissertations as useful sources of material, particularly with regard to the microwave scattering analog method for studying non-spherical scatterers as well as some theoretical methods.

Asano, S. and Yamamoto, G., 1975, Light Scattering by a Spheroidal Particle, *Appl. Opt.*, *14*, 29.

Giese, R. H., Weiss, K., Zerull, R. H., and Ono, T., 1978, Large Fluffy Particles: A Possible Explanation of the Optical Properties of Interplanetary Dust, *Astron. and Astrophys.*, *65*, 265.

Greenberg, J. M., 1978, Interstellar Dust *in* "Cosmic Dust," Ed., A. J. M. McDonnell, Wiley, London, 187.

Greenberg, J. M., 1968, Interstellar Dust *in* "Stars and Stellar Systems, Vol VII: Nebulae and Interstellar Matter," Eds., B. M. Middlehurst and L. H. Aller, University of Chicago Press, 221.

Greenberg, J. M., 1960, Scattering by Nonspherical Particles, *J. Appl. Phys.*, *31*, 82.

Jordan, E. C., 1963, Editor, "Electromagnetic Theory and Antennas," MacMillan, N. Y.

Kerker, M., 1969, "The Scattering of Light and Other Electromagnetic Radiation," Academic Press, N. Y.

Kerker, M., 1963, Editor, "Electromagnetic Scattering," Pergamon, N. Y.

Libelo, L. F., Jr., 1964, "Scattering by Axially Symmetric Nonspherical Particles Whose Size is of the Order of the Wavelength," Ph.D. thesis, Rensselaer Polytechnic Institute, Troy, N. Y.

Lind, A. C., 1966, "Resonance Scattering by Finite Circular Cylinders," Ph.D. thesis, Rensselaer Polytechnic Institute, Troy, N. Y.

Montroll, E. W. and Hart, R. W., 1951, Scattering of Plane Waves by Soft Obstacles II, *J. Appl. Phys.*, *22*, 1278.

Pedersen, J. C., 1965, "Scalar Wave Scattering by Finite Cylinders,"
  Ph.D. thesis, Rensselaer Polytechnic Institute, Troy, N. Y.
Pedersen, N. E., 1964, "A Microwave Analog to the Scattering of
  Light by Nonspherical Particles," Ph.D. thesis, Rensselaer
  Polytechnic Institute, Troy, N. Y.
Reilly, E. D., Jr., 1969, "Resonance Scattering from Inhomogeneous
  Nonspherical Targets," Ph.D. thesis, Rensselaer Polytechnic
  Institute, Troy, N. Y.
Rowell, R. L. and Stein, R. S., 1967, Editors, "Electromagnetic
  Scattering," Gordon and Breach, N. Y.
Ruck, G. T., Barrick, D. E., Stuart, W. D., and Krichbaum, C. K.,
  1970, "Radar Cross Section Handbook," Plenum Press, N. Y.
Saxon, D. S., 1955, Tensor Scattering Matrix for the Electromagnetic
  Field, *Phys. Rev.*, *100*, 1771.
Uzunoglu, N. K. and Holt, A. R., 1977, The Scattering of Electro-
  magnetic Radiation from Dielectric Scatterers, *J. Phys. A*, *10*, 413.
van de Hulst, H. C., 1957, "Light Scattering by Small Particles,"
  Wiley, N. Y.
Wang, R. T., 1968, "Electromagnetic Scattering by Spheres with
  Anisotropic Refractive Indices," Ph.D. thesis, Rensselaer Poly-
  technic Institute, Troy, N. Y.
Waterman, P. C., 1965, Matrix Formulation of Electromagnetic Scatter-
  ing, *Proc. IEEE*, *53*, 805.

SENSING ICE CLOUDS FROM SATELLITES

James T. Bunting

Air Force Geophysics Laboratory

Hanscom AFB, Massachusetts   01731

ABSTRACT

The Meteorology Division of the Air Force Geophysics Laboratory seeks improved techniques to detect cloud properties from satellites. Radiative transfer models for clouds are important tools for technique development; however, their applications to ice clouds are highly tentative due to limited knowledge of light scattering in ice clouds. The problem exists for terrestrial radiation in the IR, as well as scattered sunlight in the visible and near-IR.

1.   INTRODUCTION

When a satellite looks down over a cloudy area, it senses radiation energy which has been reflected, emitted or attenuated by the upper parts of clouds. Since the upper parts of clouds tend to be the coldest parts of clouds, the satellite is more likely to view ice particles than is a person or instrument on the earth looking up. It is obvious that an understanding of scattering by ice particles is a vital concern of anyone trying to explain the radiation which a satellite senses over cloudy atmospheres or trying to infer the properties of the clouds underneath. Particle sampling reveals a variety of shapes of ice crystals in the atmosphere so bewildering that few investigators have given them formal treatment for radiation scattering. All calculations of radiative transfer through ice clouds will have some uncertainty until measurements and calculations of phase functions have been made. This workshop represents an important step toward these goals.

25

How frequently satellites view ice clouds can be demonstrated by
means of a satellite picture. Figure 1 is a GOES-East IR image
processed by NOAA/NESS. A complicated grey scale is indicated on the
right. The image greyshades become lighter as the temperature observed by the 10.5-12.5 μm radiometer decreases to -31°C. Lower
temperatures are enhanced by contours of grey, black and white
(segments 4 through 9 of the greyscale). NESS has found that this
greyscale enhances thunderstorm identification in IR images. In
the present context, the contours can be used to identify cloudy
areas where ice crystals are almost surely present since the radiometer indicates temperatures less than -32°C and supercooled water
clouds are not expected. The area of ice clouds is underestimated
by the contoured area due to frozen clouds between 0 and -31°C and
semi-transparent cirrus with cloud temperatures less than -32°C but
with higher radiometric temperatures. Even though they may underestimate the areas of ice clouds, the contoured areas are extensive
over the central U.S.A. and Southern Canada. After viewing these
pictures from day-to-day, my subjective conclusion is that the
satellite looks at ice about half the time it looks at clouds. Here
indeed is fertile ground for applications of better knowledge of
scattering by irregular particles.

The following sections describe first some existing and planned
applications of cloud sensing from satellites and then some impli-
cations of aircraft measurements of ice particles in clouds viewed
by satellites.

2.   SATELLITE SENSING APPLICATIONS

The Air Force uses weather satellite data to detect clouds and
as input to cloud forecasting models. A detailed technical report by
Fye (1978) describes the Air Force Global Weather Central (AFGWC)
automated cloud analysis program (3DNEPH) which produces high resolution, three-dimensional analyses of clouds over the entire globe.
All clouds, including both ice and water clouds, are sought. Both
visible and IR imagery from polar-orbiting satellites are used for
cloud detection. The cloud analysis from the 3DNEPH, along with
three-dimensional trajectories calculated using forecast winds, are
used to forecast cloudiness at various times following the satellite
passes. The day-to-day cloud analyses also form the basis of design
climatologies.

The Air Force Geophysics Laboratory is developing techniques to
improve the utility of existing visible and IR data and also to
utilize new data from polar-orbiting satellites at near-IR and microwave frequencies. Moreover, data from geostationary satellites are
collected and used to develop mesoscale cloud analysis and forecasting
techniques which benefit from the short time interval (30 min.)
between successive satellite observations.

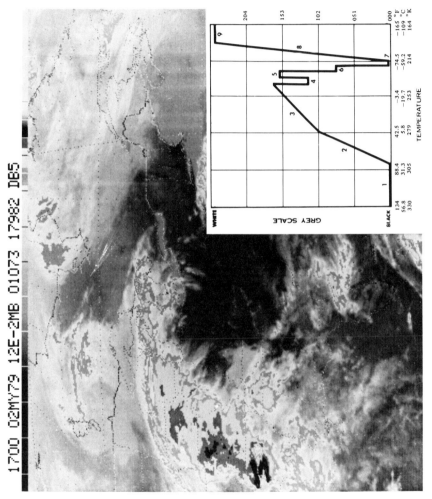

FIG. 1.   An IR picture of the Eastern USA and Canada taken by the GOES-East satellite at 1700Z, 2 May 1979.   Temperatures below -31°C are contoured according to the greyscale in the lower right and these contoured areas are expected to be ice clouds.

In my opinion, improved phase functions for ice particles could primarily improve satellite sensing of clouds in two ways: by improving bidirectional reflectance models for normalizing channels with reflected sunlight to some standard viewing geometry and also by providing more realistic models for radiative transfer in ice clouds. An example of cloud scattering variation is shown in Figure 2 which is a display of 13 orbits of NOAA 4 visible satellite data mapped into a polar stereographic projection. NOAA 4 has a morning (0900 Local Time) descending sun-synchronous orbit. At high latitudes the boundaries of orbital swaths are apparent in cloudy areas. Although the data are normalized for solar illumination, they are not corrected for changes in scattering. Near the boundaries at high

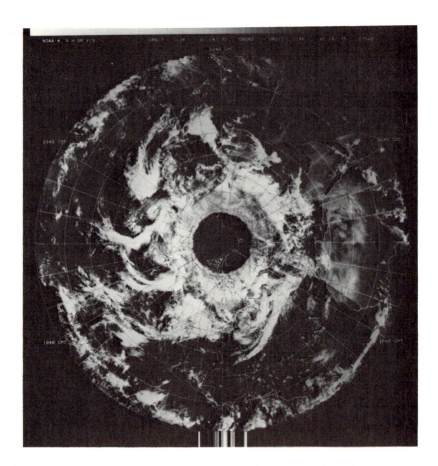

FIG. 2.   Thirteen consecutive orbits of NOAA-4 visible imagery mapped into a polar stereographic projection of the Northern Hemisphere. The boundaries between orbital swaths are easy to see at high latitudes and they represent the difference between forward scattering and backscattering.

latitudes, the satellite is looking at backscattered sunlight on the
eastern side of the boundary, and, two hours later, it looks at
forward scattered sunlight on the western side of the boundary.
Models of bidirectional reflectance have been published by Raschke
et al. (1973) and have been applied to radiation budget studies.
These models appear to be based on observations over liquid clouds
such as stratus and their applicability to ice clouds is not known.
Although particle shape is only one of many considerations that will
make it difficult to implement operational bidirectional reflection
models for remote sensing, the potential payoff is large since the
models will extend the automated analysis of visible data to
morning/evening orbits for sun-synchronous satellites and to more
daylight hours for geostationary satellites.

Radiative transfer models are used to explain what satellites
are already sensing and to predict what new channels will sense.
Their present applications to scattering in ice clouds are highly
tentative and improvements would be useful.

3. RADIOMETRIC CONSIDERATIONS

Scattering requirements are defined not only by the shapes of
ice particles, which are discussed in Section 4, but also by the
characteristics of sensors on satellites. Table 1 identifies
spectral channels which are now flying or are scheduled to in the
future.

The table includes imagery channels of DMSP as well as TIROS N
satellites. In order to be brief, Nimbus satellite channels as well
as all sounding channels have not been included even though some of
the channels are strongly influenced by clouds. The importance of
scattering in the visible and near IR channels need not be emphasized.
There is considerable absorption by ice particles in the IR channels
and many investigators are tempted to make the risky assumption that
clouds behave as blackbodies in the IR channels. However, Liou
(1974) has demonstrated the importance of ice particle scattering
in the IR window region and the difficulties which follow if the
particles are treated as equivalent spheres. Table 1 also includes
some microwave channels which might appear to be of little interest
to this workshop if the axiom that "ice clouds have no effect on
microwaves" is invoked. However, microwaves are scattered by ice
particles even if not absorbed. They are absorbed by melting snowflakes, and future satellites will be capable of measurements at
shorter wavelengths at which scattering should be greater. Finally,
for all wavelengths except the microwaves, the bandwidth of the
channel should be considered in scattering measurements or calculations.

Although the following considerations depart from the central
workshop theme of phase functions for various particle shapes, they

TABLE 1

PRESENT AND FUTURE SPECTRAL CHANNELS

Spectral region	Central wavelength or bandwidth
Visible/Near IR	0.5 - 1.0 μm
	0.55- 0.70 μm
	0.72- 1.10 μm
Near IR	1.52- 1.63 μm
IR	3.55- 3.93 μm
	8.0 -13.0 μm
	10.5 -11.5 μm
	10.5 -12.5 μm
	11.5 -12.5 μm
Microwave*	94 GHz (0.32 cm)
	37 GHz (0.81 cm)
	19 GHz (1.58 cm)

are important sources of variation in remote sensing and cannot be easily separated from variation in particle shape. The sensor fields of view at cloud level vary from less than 1 km to about 20 km. Even the smallest field of view sees an enormous volume compared to an experimental situation such as a cloud chamber so that the shapes of the clouds themselves influence the satellite measurements. Many ice clouds are thick enough that radiation sensed by satellites has undergone multiple scattering. Finally, there is inevitably some contribution by absorbing gases, aerosols, or surfaces.

4.   ICE PARTICLE OBSERVATIONS

The Air Force Geophysics Laboratory has made a series of air-craft flights underneath weather satellites (Conover and Bunting, 1977). Aircraft instrumentation measured cloud particles and pro-vided estimates of the vertical distribution of cloud and precipi-tation mass density underneath the satellite. The commonly avail-able satellite measurements of reflected sunlight at 0.5 to 0.7 μm and IR radiance at 10.5 to 12.5 μm were found to be good predictors of vertically integrated cloud properties such as total cloud mass and thickness in empirical equations (Bunting, 1978). The aircraft samples of ice particles revealed a highly variable assortment of

---

* Microwave channels have not been selected for the Defense Meteoro-logical Satellite Program (DMSP) microwave imager. The above have been suggested in earlier studies.

shapes, sizes, mass densities and number densities which, at that
time, discouraged us from using the data in radiative transfer
calculations.

Although the variability of ice particles is considerable, the
particle shapes and sizes are not random and some order can be seen
if the particle data is stratified by temperature. Figure 3 has
samples of particle cross sections as displayed by a two dimensional
optical array from Particle Measuring Systems, Inc. (Knollenberg,
1970). These data were taken at various altitudes in a flight on
8 October 76 over Vermont. The air temperature (°C) is given on the
left of each sample. The vertical bar on the left of each particle
represents a length of 1.28 mm. At 5°C, the particles are raindrops.
At 0 to -10°C, they are mostly large snow, sometimes agglomerations
of more than one flake. A stellar dendrite is circled. At -15 to
-25°C, the particles are mostly small snow and at -30 to -40°C, the
coldest part of the cloud, they are clusters of prisms known as
bullet rosettes. The data in Figure 3 represent variation in pre-
cipitation-sized particles over a depth of about 10 km of cloud and
are typical of precipitating stratiform clouds at temperate latitudes. The following generalizations can be made for such clouds
based on the aircraft sampling experience. First, particle size
and cloud mass density both tend to decrease with temperature. The
physical basis for this observation is the fact that water vapor
density, the source of growth for the particles, decreases with
temperature. Second, very few regularly shaped ice particles with
a pleasing geometrical form are to be found. The particle categories

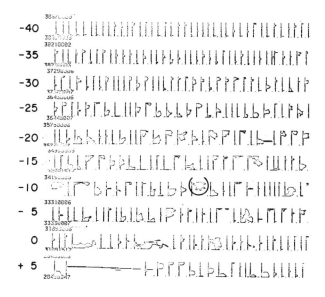

FIG. 3.   Cross sections of particles sampled by the 2-D PMS probe
during an aircraft ascent on 8 October 1976.

such as bullet rosettes or small snow are to some extent conveniences. Fragments of particles are frequently present and it is often diffi- cult to determine the altitudes at which the categories change. Third, the smallest ice particles of about 30 µm or less were not sampled as well as larger particles. Finally, in most practical situations, there are sizable uncertainties to estimates of quantitative prop- erties such as size distributions or mass densities. These uncer- tainties are partly due to the irregular particle shapes and partly to the fact that the aircraft samples relatively few particles from the cloud.

The data shown in Figure 3 along with data from other sampling instruments on the aircraft were converted to altitude profiles of cloud mass density shown in Figure 4. The open circles represent data taken before a satellite pass while the aircraft ascended and the shaded circles represent data taken after the satellite pass while the aircraft descended. In Figure 4, the mass density tends to decrease from about 0.5 g/m^3 at 2 km to less than 0.001 g/m^3 above 10 km. The general decrease of mass with increasing altitude and decreasing temperature is complicated by considerable vertical structure which varied from ascent to descent.

The satellite measurements of IR and visible appeared to be sensitive to cloud properties to considerable depths within the cloud so that it would be difficult to ignore the scattering prop- erties of any of the ice particle classes. The IR sensor at 10.5 to 12.5 µm measured an equivalent temperature of −31°C which corres- ponds to an altitude of 8.6 km. Since the IR is sensing particles above and below that altitude, it is looking into the upper 2 or 3 km of the cloud, which consists of bullet rosettes and small snow. The visible sensor at 0.5 to 0.7 µm indicated a reflectivity of slightly over 60%, which suggests that sunlight is penetrating the cloud fairly well and that significant scattering takes place at all alti- tudes within the ice cloud. Microwave channels would readily see down to the melting snow and rain in the lowest parts of the clouds.

5. CONCLUSIONS

It appears to be much easier to define the spectral bandwidths of satellite channels than it is to define ice particle shapes and scattering properties. Some particle sizes and shapes are not easily measured by aircraft in clouds. Improved scattering models for ice particles will be necessary before cloud particle observations can be related to satellite radiation observations within reasonable limits of error. Even with improved scattering models, however, the varia- bility of ice crystals in clouds is so great that field experiments to relate cloud and satellite measurements should continue whenever possible.

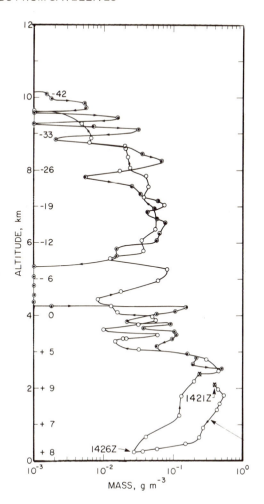

FIG. 4.   Vertical profiles of mass for particles greater than 50 μm
in diameter during ascent and descent on 8 October 1976.   Temperatures in °C are shown on the left.   Points are 1-D PMS probe
measurements.   Points with crosses (X) are averages over more than
one sampling run.

Question (VONNEGUT):   Have your satellite observations of ice clouds
revealed any cases of specular reflection?   I have observed such cases
from aircraft.

Answer (BUNTING):   I assume that specular reflection requires ice
crystals to be flat and horizontally oriented.   We observe ice clouds
with crystals like these in our own aircraft sampling of ice parti-

cles, and specular reflection of sunlight is reasonable to expect. The satellite data we used was, however, restricted in viewing and scattering angles so that specular reflection wasn't observed. It would be a useful indicator for ice clouds under satellites for those small areas where the viewing geometry is right, provided that it could be distinguished from ocean sunglint.

Question (GREENBERG):   Would you be able to get extra information on discriminating between ice and water clouds by comparing measurements made at wavelengths of 2.7 and 3.1 μm?

Answer (BUNTING):   We do not make that comparison, however, we do use a channel at 1.6 μm which was chosen because water and ice look different there and can be distinguished, and because it is a fairly good window for seeing down into the atmosphere.

Question (GOEDECKE):   What is the orientation of your particle counter? The reason I ask is that, if for example, snowflakes have a strong preferential orientation, presumably horizontal, and if your two-dimensional imaging views only the horizontal directions, the shadow would appear to be a straight line for most flakes. If the flakes tilt only a little away from the horizontal, the shadow would appear considerably smaller than the actual area of the flake. Even if you view one horizontal and one vertical direction, how do you infer the other projection and thereby, the twice-dimensional form?

Answer (BUNTING):   The laser beam is vertical so that flat or long particles falling with their long axes horizontal will be imaged most favorably. Empirical adjustments have been used to adjust for orientations out of the plane of the sensor.

REFERENCES

Bunting, J. T., 1978, *in* "Preprints of 3rd Conf. on Atmospheric Radiation," Davis, CA, June 1978, Amer. Meteor, Soc.

Conover, J. H., and Bunting, J. T., 1977, "Estimates from Satellites of Weather Erosion Parameters for Reentry Systems," AFGL-TR-77-0260, 85.

Fye, F. K., 1978, "The AFGWC Automated Cloud Analysis Model," AFGWC Tech. Memo 78-002, 97.

Knollenberg, R. G., 1970, *J. Appl. Meteor.*, 9, 86-103.

Liou, K. N., 1974, *J. Atmos. Sci.*, 31, 522-532.

Raschke, E., VonderHaar, T. H., Pasternak, M., and Bandeen, W. R., 1973, "The Radiation Balance of the Earth-Atmosphere System from Nimbus 3 Radiation Measurements," NASA TN D-7249, 71.

LIDAR VISIBILITY MEASUREMENTS

Eugene Y. Moroz

Air Force Geophysics Laboratory, Meteorology Division

Hanscom Air Force Base, Massachusetts   01731

ABSTRACT

The Air Force requires information on light scattering by atmospheric particles in order to determine visibility for aircraft operations at airfields. The multiple scattering process is one facet that comes into strong play in the measurement of visibility in category II and III conditions. Experimental lidar slant visual range measuring systems are described.

For a number of years experimenters have been investigating the possibility of determining the atmospheric extinction coefficient from lidar atmospheric backscatter observations. Much of the effort has been directed toward the development of an operational slant range visibility measuring lidar system for routine airfield use. A brief survey is presented of lidar programs involved in this area including the programs sponsored by the Air Force Geophysics Laboratory (AFGL).

An early effort was conducted by United Aircraft Corporate Systems under the direction of Hubbard (Hubbard *et al.*, 1965). An experimental coaxial lidar system was fabricated which used a ruby laser source. In November, 1965, measurements were obtained at Rentschler Field, East Hartford, Connecticut during three fog episodes. The data were evaluated using a slope technique. The results appeared to confirm the theory of the technique showing that the atmospheric scattering coefficient and transmission could be determined from a measurement of lidar backscatter.

Brown (1967) derived a number of relationships between characteristic lidar backscatter signatures and the extinction coefficient.

These signatures included the logarithmic slope, the distance to the peak return and the width of the return at an arbitrary power level. Brown showed that atmospheric transmittances derived from signatures from a non-coaxial lidar compared well with the values obtained with a standard transmissometer. An important conclusion was that in very dense fog the effects of multiple scattering influenced the quantitative results.

Another relationship that was evolved from the lidar backscatter signature study is used in the so-called ratio technique. A solution of the lidar equation is obtained by taking power ratio measurements of the backscatter return from two adjacent ranges. The technique, which was developed for use with low-powered, high repetition rate laser sources such as galluim arsenide, was demonstrated by Brown (1973) and Lifsitz (1974).

There has been interest in the Raman scattering technique as a means to solve the lidar equation. The volume backscatter scatter coefficient for Raman scattering depends only on the Raman cross section of the molecule and its number density. In the lower atmosphere the density of atmospheric nitrogen is constant. Therefore, a measurement of Raman scattering from nitrogen should give a direct determination of atmospheric transmittance as a function of range. Experimental field measurements by Leonard (1978) demonstrated the feasibility of the Raman method.

Subsequent work at Stanford Research Institute under Collis (1970) furthered the development of the lidar slope technique. A series of field experiments were conducted. The final experiment which was performed at Travis AFB, California during January 1973, was designed to evaluate precisely the lidar slope technique (Viezee *et al.*, 1973). Visual range computed from lidar backscatter observations showed high linear correlation to visual range measured simultaneously by transmissometers. However, the lidar derived values consistently over-predicted the transmissometer measurements. As with Brown's results, it was concluded that at low visibilities a large part of the overprediction must be attributed to large forward scatter and multiple scatter.

Stewart (1976) evolved a novel "analog zone" lidar technique for determining atmospheric transmission. A diagram of the lidar is shown in Figure 1. The transmission is obtained from the square root of the measurement of the ratio of two integrations over the path of interest. A major feature of the technique is that no *a priori* information regarding the extinction coefficient and the phase function is required. The extinction coefficient may vary along the measurement path, the only assumption being a constant phase function. The Raytheon Co., which was contracted to fabricate a system using this technique for AFGL, was unable to provide an operable lidar. Work on the system is continuing in-house at AFGL.

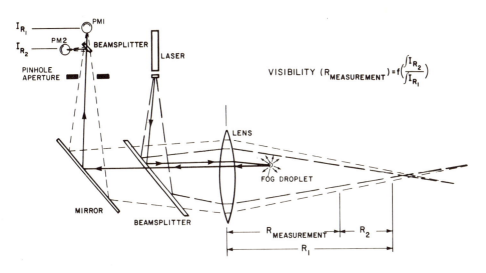

FIG. 1.   Lidar slant visual range measuring system.

     In conclusion, a number of techniques and systems for the mea-
surement of visibility have been investigated and their feasibility
has been successfully demonstrated to some degree. However, after
more than fifteen years of effort, there still remain a number of
areas that need to be resolved before an operational lidar visibility
lidar can be developed. These include the effect of multiple scatter,
measurements in patchy fog conditions, laser beam penetration, timely
measurements, processing in real time and eye safe operation.

Question (JUISTO): Have you found in your data any reason to suspect
that there is a direct relation between horizontal and slant path
visibility measurements?

Answer (MOROZ):   In general we find an advection fog to be "homogeneous" in the horizontal but not in the vertical direction. In a
falling stratus that eventually becomes a ground fog, the same is
true. The vertical extinction gradient can be in opposite directions
for these two situations.

                         REFERENCES

Brown, R. T., Jr., 1973, *J. Appl. Meteor.*, 12, 698-708.
Brown, R. T., Jr., 1967, "Backscatter Signature Studies for Hori-

zontal and Slant Range Visibility," Final Report, May, 1967, AD 659 469.

Collis, R. T. H., Viezee, W., Uthe, E. E., and Oblanas, J., 1970, "Visibility Measurement for Aircraft Landing Operations," Final Report AFCRL-70-0598, September, 1970.

Hubbard, R. G., Bukowski, C. F., Waters, N. L., and Holmes, R. E., 1965, "Slant Visual Range Feasibility Study," Interim Report, WSC E-55, December, 1965.

Leonard, D. A., 1973, "An Experimental Field Study of a Raman Lidar Transmissometer," IEEE/OSA Conference on Laser Engineering and Applications, Digest of Technical Papers, 36-37.

Lifsitz, J. R., 1974, "The Measurement of Atmospheric Visibility with Lidar," Final Report, FAA-RD-74-29, March, 1974.

Stewart, H. S., Shuler, M. P., Jr., and Brouwer, W., 1976, "Single Ended Transmissometer Using the Analog Zone Principle," HSS-TD-043, August, 1976.

Viezee, W., Oblanas, J., and Collis, R. T. H., 1973, "Evaluation of the Lidar Technique of Determining Slant Range Visibility for Aircraft Landing Operations," Final Report - Part II AFCRL-TR-73-0708, November, 1973.

NON-SPHERICAL PARTICLE SCATTERING: AIR FORCE APPLICATIONS

Eric P. Shettle

Air Force Geophysics Laboratory

Hanscom Air Force Base, Massachusetts 01731

ABSTRACT

Particles of primary interest range from natural atmospheric aerosols through cloud droplets and meteoric dust. Problem areas include distinguishing between ice and water clouds, and the effect of irregularly shaped particles on the measurement of aerosol properties using scattered light (such as lidar measurements or optical particle counters).

1. INTRODUCTION

As a representative of the user community, I have been requested to describe our needs for information on the optical properties of irregularly shaped particles and to indicate the size distribution and refractive index of the particles we are interested in. This paper will briefly describe the Air Force's primary interests and needs in the area of light scattering by non-spherical particles.

Any optical or IR system that operates in or above the atmosphere must look through the atmosphere or against a background of either the atmosphere or space. Therefore, both the optical properties of the environment and how they will effect the operational performance of these systems must be known.

At AFGL, we carry out experimental and theoretical research on the optical/IR properties of the aerospace environment and attempt to model and predict its effects on the propagation of optical and IR radiation. These effects include transmission losses, contrast reduction and backgrounds of scattered and/or emitted radiation against which a signal or object must be seen.

The types of irregularly shaped particles of interest cover a wide range from "naturally" occuring atmospheric haze particulates of ice clouds to interplanetary dust particles. Within the Air Force there is some interest in the optical properties of the particles in the battlefield environment, which include artificial clouds, although this is primarily the responsibility of the Army.

The remainder of this paper will describe what we know about the different types of aerosols in greater detail and how uncertainties in the optical properties impact the applications.

2. TYPES OF NON-SPHERICAL PARTICLES

2.1 Natural Atmospheric Aerosols

The atmospheric aerosols originate from a number of different sources. Much of the atmospheric aerosols are derived from material on the earth's surface such as soil, salts, or pollen. Most of the remainder is produced by chemical or photochemical processes within the atmosphere. The prime example of this is the stratospheric sulfate aerosols (Turco *et al.*, 1979 and Toon *et al.*, 1979). There is also some interplanetary dust settling downward through the earth's atmosphere, but this is a negligible component, except in the upper atmosphere above 25 to 30 kilometers (Rosen, 1969). Some typical aerosol refractive indexes for the lower atmosphere are shown as a function of wavelength (Volz, 1972, 1973, 1979; Twitty and Weinman, 1971) in Figures 1 and 2.

The size of the optically important particles ranges from about 0.02 to 50 microns. Typical size distributions are shown in Figure 3. There is some uncertainty introduced into the measurement of the particle size by many of the optical types of measurement which generally assign the size of a particle as the size of an equivalent sphere whose scattering properties are similar to those measured. The problem is what happens when we use this "equivalent size" to calculate some other optical property such as the total cross section or the scattering at a different angle and wavelength.

Another problem is that many of the techniques for remote sensing of the atmospheric aerosols, such as lidar, rely on particle optical properties which are sensitive to the assumption of spherical particles, such as the backscatter cross section. (Some of these problems are discussed by Moroz (1980) in another paper at this Workshop).

We are also interested in various types of liquid aerosols such as cloud and fog droplets. However, these are small enough that deviations from sphericity can be neglected. Of an intermediate nature are haze particles at high relative humidities (greater than 75 to 90%) where a spherical shell of water may cover an irregularly shaped

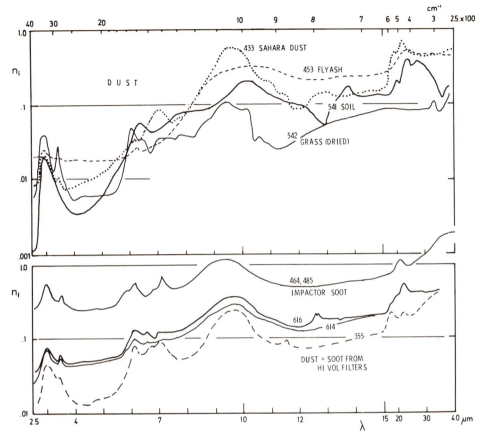

FIG. 1.   Measured refractive indexes for various aerosol samples
from the lower atmosphere.

insoluble nucleus. The question of interest for this workshop is:
under what conditions can the non-spherical nature of the core be
neglected?

2.2   Ice Clouds

Ice clouds include Cirrus, Nacreous and Noctilucent clouds. The
refractive index of ice is well known (Irvine and Pollack, 1968;
Schaaf and Williams, 1973) and is shown in Figure 4. The size and
shape of the ice crystals depends on the temperature and to a lesser
extent on the degree of super-saturation when the crystal forms (see
Byers, 1965 or Mason, 1957). The shapes may include some irregular
features but are generally based on a hexagonal form, ranging from
needles and columns to plates. The sizes vary from about 10 to 200

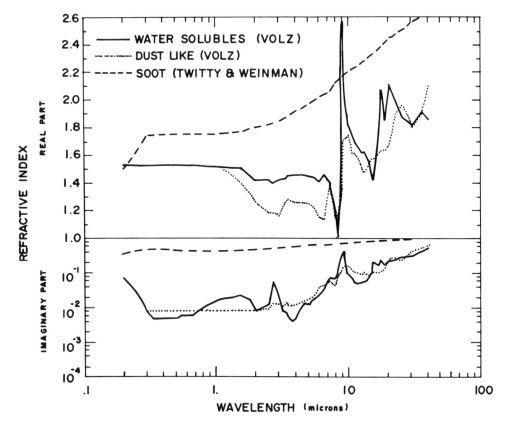

FIG. 2.   Model refractive indexes for typical atmospheric aerosols.

microns.   The properties of ice clouds are discussed in greater detail
by Bunting (1980) in this volume.

2.3   Interplanetary Dust

As noted previously, the major component of the upper atmospheric
aerosols is interplanetary or meteoric dust settling downward through
the atmosphere.   Figure 4 shows the refractive index for a mixture
of chondrite dust (Shettle and Volz, 1976) which represents the major
type of meteorite incident on the earth.   It should be noted that
the refractive index is based on fitting a damped harmonic oscillator
dispersion model to reflectance measurements between 2.5 and 40
microns, and the values for shorter wavelengths are just an extrapo-
lation.

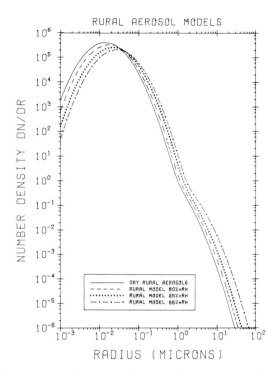

FIG. 3.   Typical aerosol size distribution as a function of relative humidity (from Shettle and Fenn, 1979).

At AFGL there is a research effort to measure and model the background radiance from the zodiacal light (Murdock and Price, 1978, 1979) as depicted in Figure 5. One model (solid line) uses a size distribution similar to Giese and Grun (1976) with spherical particles and the refractive index of obsidian (Pollack *et al.*, 1973). The dashed line shows the effect of artificially increasing the imaginary part of the refractive index. The model shows good agreement with the measurements of Nishimura (1973) and of Hoffman *et al.* (1973) if the number densities are a factor of 10 less than Giese and Grun (1976). The meteoric dust refractive index of Shettle and Volz (1976) was also used instead of obsidian in the zodiacal light model, but the results did not agree as well with the observations (Murdock, 1979).

No reasonable refractive index and size distribution seem to be able to reproduce the spectral data at all angles, including the Gegenschein, using Mie calculations with spherical particles. However, the particles are known to be irregularly shaped.

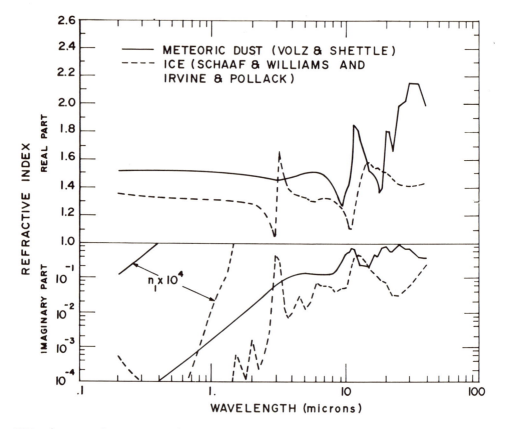

FIG. 4.   Refractive index vs. wavelength for ice and meteoric dust.

3.   APPLICATIONS

For transmission modeling the primary contribution of the
aerosols is in terms of their total extinction cross section. Figure
6 shows the spectral dependence for several different aerosol models
(Selby *et al.*, 1976). The uncertainties introduced by a lack of
knowledge of the exact shape of particles is relatively small compared
to the uncertainty in the size distribution.

The effects of nonsphericity are most significant when one is
observing scattered radiation, especially singly scattered radiation.
At AFGL we have been studying the scattering properties of the aerosols and have recently built a dual wavelength IR polar nephelometer
to make *in situ* measurements of the phase function of the atmospheric
aerosols at 1.06 and 10.59 microns. Figure 7 shows the results of a
test of the 1.06 micron system in a clean room which was filled with
a monodispersion of latex spheres with a diameter of 0.945 microns.

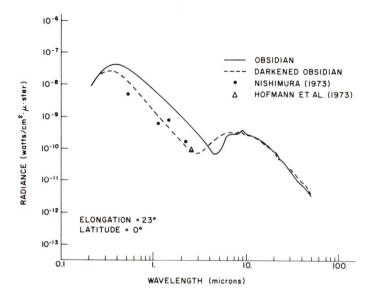

FIG. 5.   Comparison of zodiacal light radiance model with observation.

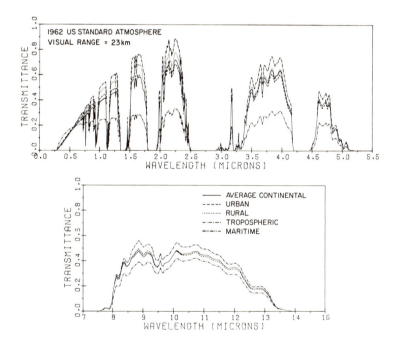

FIG. 6.   Atmospheric transmittance vs. wavelength for 10 km path at
sea level for various aerosol models.

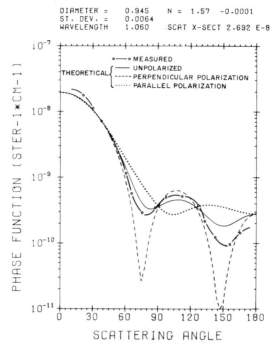

FIG. 7.   Comparison of measured and theoretical angular scattering function for a wavelength of 1.06 microns with polar nephelometer.

We have recently begun taking outside measurements of the ambient aerosols with the system but the data has not been reduced yet.

Question (SCHAEFER):   What angular scattering functions have you used in your zodiacal light models?

Answer (SHETTLE):   If I did not make it clear in my presentation, let me say that the zodiacal light models are Dr. Tom Murdock's; my contribution was to provide him with a Mie code to use. The particles were thus treated as spheres.

Question (SRIVASTAVA): What was your theoretical model to study the scattering from meteoric dust grains in the stratosphere and what are its limitations compared to actual grain shape?

Answer (SHETTLE):   Our aerosol model for the upper atmosphere (above 40 to 50 km) consisted of spherical meteoric dust. The only detectable effects of this component are in terms of scattered sunlight; the transmission for tangent paths above 40 to 50 km is better than 99.9%. Uncertainties in the number and size distributions are more important than the shape.

Question (WEISS): Did you investigate the linear polarization of the scattered light? Most of the zodiacal light models based on Mie calculations using realistic size distribution and material cannot reproduce the measured linear polarization.

Answer (SHETTLE): The light scattering measurements that I described were a test of a system designed to make *in situ* measurements of the natural atmospheric aerosol at the Earth's surface — not the interplanetary dust. The 1.06 μm system uses unpolarized light; the 10.6 system uses polarized light. While Murdock's zodiacal model uses spherical particles, we are aware of the need to consider irregular shapes. While the model reproduces the spectral dependence at 23°, it does not have the proper angular variation nor does it reproduce the Gegenschein.

REFERENCES

Bunting, J., 1980, this volume.

Byers, H. R., 1965, "Elements of Cloud Physics," Univ. of Chicago Press, p. 191.

Giese, R. H. and Grun, E., 1976, *in* "Interplanetary Dust and Zodiacal Light," H. Elsasser and H. Fechtig, Eds., Spring-Verlag, Berlin, p. 135-139.

Hofmann, W., Lemke, D., Thum, C. and Fahrbach, U., 1973, *Nature Phys. Sci.*, *243*, 140.

Irvine, W. M. and Pollack, J. B., 1968, *Icarus*, *8*, 324-360.

Mason, B. J., 1957, "The Physics of Clouds," Oxford Univ. Press.

Moroz, E. Y., 1980, this volume.

Murdock, T. and Price, S., 1978, "Infrared Emission of the Interplanetary Dust Cloud," presented at the 152nd meeting of the Amer. Astron. Soc. 26-28 June, 1978, Madison, WI (Abstract in *Bull. Amer. Astron. So.*, *10*, 459, 1978).

Murdock, T., 1979, private communication.

Nishimura, T., 1973, *Publ. Astron. Soc. Japan*, *25*, 375-384.

Pollack, J. B., Toon, O. B., and Khare, B. N., 1973, *Icarus*, *19*, 372-389.

Rosen, J. N., 1969, *Space Sci. Rev.*, *9*, 58-89.

Schaaf, J. B. and Williams, D., 1973, *J. Opt. Soc. Am.*, *63*, 726-732.

Selby, J. E. A., Shettle, E. P., and McClatchey, R. A., 1976, "Atmospheric Transmittance From 0.25 to 28.5 μm: Supplement LOWTRAN 3B," AFGL-TR-76-0258 (1 Nov., 1976).

Shettle, E. P. and Fenn, R. W., 1979, "Models for the Aerosols of the Lower Atmosphere and the Effects of Thermal Variations on their Optical Properties," AFGL-TR-79-0214, Sept. 17.

Shettle, E. P. and Volz, F. E., 1976, "Atmospheric Aerosols: Their

Optical Properties and Effects," 13-15 Dec., 1976, Williamsburg,
VA, Opt. Soc. Amer. & NASA LRC, NASA-CP-2004, pp. MC14-1 to MC14-4.
Toon, O. B., Turco, R. P., Hamil, P., Kiang, C. S., and Whitten, R.
E., 1979, *J. Atmos. Sci.*, *36*, 718-736.
Turco, R. P., Hamil, P., Toon, O. B., Whitten, R. C. and Kiang, C. S.,
1979, *J. Atmos. Sci.*, *36*, 699-717.
Twitty, J. J. and Weinman, J. A., 1971, *Journ. Appl. Meteor.*, *10*,
725.
Volz, F., 1972, *J. Geophy. Res.*, *77*, 1017-1031.
Volz, F., 1972, *Appl. Opt.*, *11*, 755-759.
Volz, F., 1973, *Appl. Opt.*, *12*, 564-568.
Volz, F., 1979, private communication.

PARTICLES PRODUCING STRONG EXTINCTION IN THE INFRARED

Janon Embury

U.S. Army Chemical Systems Laboratory

Aberdeen Proving Ground, Maryland   21010

ABSTRACT

A summary of particle shapes, sizes and compositions which produce strong extinction per unit mass are presented. These particles are of considerable interest to the Army Smoke Program at Chemical Systems Laboratory.

Over the last several years the Army has had a growing interest in producing a screening smoke in the infrared, particularly the 3-5 microns and 7-14 microns atmospheric windows. Eventually this interest will expand into the millimeter wave region. We want to understand what improvements over conventional screening smokes are possible without restricting constituent particle size, shape, or complex refractive index. To do this in a brute force and exhaustive way would mean understanding many aspects of scattering and absorption by virtually every imaginable type of particle. Clearly some guidelines are needed to reduce the scope of such a search.

In smoke clouds most forms of small particles will be randomly oriented so we would encourage that scattering and absorption calculations and measurements be carried out not just for some preferred particle orientation but rather for an ensemble of random orientations. When characterizing the interaction of light with a particle we need to know first and foremost the extinction by particles having a complex refractive index. Secondly, we would like information concerning the albedo and phase function in order to know how much scattered light passes through the cloud. If full information about

the phase function is difficult to ascertain, it is sometimes sufficient to describe only the asymmetry factor and small angle forward
scattering. Finally, and of least importance at this time, knowledge about the polarization of scattered light is desired. Screening smoke clouds typically have a concentration on the order of 10^6
particles cm^{-3} which could conceivably perturb single scattering and
force the researcher to consider particle-particle interactions.
This can make scatter measurements bothersome since most particle
sizing instruments that use scattered light develop coincidence
counting problems at these concentrations. Once a particle is born
into existence we want to be able to measure its size, shape, and
refractive index. The researcher must establish the information requirements needed to invert scattered light data, and the experimen-
talist will have to develop compatable methods to create this data.

We anticipate that when a particle providing exceptional screening is found or predicted it will be necessary to engineer size,
shape, and refractive index. It is unlikely that any natural or
anthropogenic aerosol or any commercially available particulate
materials, pigments for example, will produce the strong extinction
we seek. Our needs in the Army smoke program are not unique; solar
energy researchers are interested in finding solar selective coatings
for solar absorbing panels composed of particulate material which
absorbs solar wavelengths strongly and emits infrared wavelengths
weakly.

There exist a great variety of arbitrarily shaped non-spherical
particles. Particles can be homogeneous or non-homogeneous, regular
non-spherical or irregular non-spherical, elongated or flattened,
convex or concave, have edges and vertices or have none. Homogeneous particles have a constant internal refractive index while non-
homogeneous particles have a spacially variable internal refractive
index. Irregular non-spherical particles superimpose some degree of
random roughness on the smooth surfaces and edges of a regular non-
spherical particle. Examples of regular convex non-spherical particles with increasing numbers of edges and vertices are cylinders,
tetrahedra and octahedra. Examples of regular convex non-spherical
particles without edges and vertices are spheroids, ellipsoids and
particles with more than three lobes. Examples of non-homogeneous
particles are layered structures, foams and clusters or aggregates
of particles which scatter and absorb radiation as a single unit.

Presently the particles we have found that produce the greatest
extinction over the infrared atmospheric windows are small absorbing
particles, flakes, hollow particles, foamed or puffed particles and
particles producing optical resonance of some kind. This last type
of particle will be discussed in a later session. We can understand
the source of high extinction by small absorbing particles, flakes,
hollow particles and foamed particles if we look closely at the
definition of the extinction coefficient, α, of a particle

$$\alpha = \frac{\langle \text{Geometric Cross Section} \times \text{Extinction Efficiency} \rangle}{\text{Weight}}$$

where the brackets indicate an average over the ensemble of random orientations. Particles exhibiting optical resonance produce high extinction by increasing the extinction efficiency. On the other hand, the four types of particles just mentioned increase extinction by increasing the ratio of geometric cross section to weight.

Small absorbing spheres have the same geometric cross section for all orientations and an extinction coefficient given by

$$\alpha = \frac{\pi r^2 Q}{\rho^4/3 \pi r^3} = \frac{3Q}{4\rho r}$$

where r = radius, ρ = density, and Q = extinction efficiency.

The ratio of geometric cross section to weight is inversely proportional to radius so that as the radius becomes smaller this ratio grows and extinction grows. It turns out that there is a limit imposed on this growth by the extinction efficiency. For a non-absorbing sphere much larger than a wavelength the extinction is inversely proportional to radius until the extinction peaks at a size comparable to the wavelength and thereafter is inversely proportional to r^3 (decreasing as the radius becomes smaller).

An absorbing sphere much larger than a wavelength also produces extinction inversely proportional to radius until the size becomes comparable to the wavelength, but thereafter extinction plateaus as the radius becomes smaller instead of decreasing as in the case of non-absorbing spheres.

The ratio of geometric cross section to weight for a metal sphere can be increased by removing the core of the sphere until the remaining shell thickness approaches the skin depth. In essence, we are reducing the weight of the sphere by hollowing it out without altering the external dimensions or optical properties. If we define the shell thickness t, then the extinction of a thin shelled hollow metal sphere is given by

$$\alpha = \frac{\pi r^2 Q}{\rho 4 \pi r^2 t} = \frac{Q}{4\rho t} \quad .$$

When the sphere size becomes large compared to wavelength, the extinction efficiency, Q, approaches a value of 2. This is true for large particles of any shape (van de Hulst, 1957). For the case of a large aluminum sphere with shell thickness 0.2 micron and density 2.7 g cm^{-3}, the extinction becomes 0.92 m^2 g^{-1}. At longer wavelengths when shell thickness becomes less than the skin depth the absorption efficiency decreases; however, the scattering efficiency increases holding the extinction efficiency at the constant value of two provided the wavelength remains small compared to sphere size.

We can generalize the discussion about hollow metal spheres to include any hollow thin shelled convex particle. The time average or, equivalently, the ensemble average geometric cross section for a convex particle is simply one fourth its surface area (van de Hulst, 1957):

<Geometric Cross Section> = Surface Area/4.

The weight of a thin walled hollow particle is

Weight = Density × Thickness × Surface Area.

The extinction coefficient for a convex thin walled particle of any shape is then given by $\alpha = Q/(4\rho t)$.

A randomly oriented thin planar flake of uniform thickness t has an average geometric cross section equal to one fourth its surface area. The weight of this flake, ignoring edge surface area becomes

Weight = Density × Thickness × Surface Area/2,

resulting in an extinction coefficient twice that of a hollow particle having the same density, wall thickness, and efficiency: $\alpha = Q/(2\rho t)$. Thus, a large planar aluminum flake with thickness 0.2 micron and an efficiency of two produces an extinction coefficient of 1.85 m^2 g^{-1} or twice what was computed earlier for large hollow aluminum spheres.

It is interesting to return to solid spheres for a moment to determine what extinction efficiency would be necessary to produce extinction equivalent to or greater than that of a large flake having an efficiency of two. When this is done we find that $Q \geq 4r/(3t)$.

When flakes become very thin, where thickness is much less than incident radiation wavelength, the only way to satisfy this inequality is with some type of resonance phenomena.

Finally, the geometric cross section of a given mass of material can be increased by foaming the particle with air. The inclusion of microvoids small compared to wavelength causes the effective refractive index to approach that of air instead of scattering radiation within the particle. The radius of the foamed or puffed sphere, r,

as a function of initial unpuffed radius, r_0, and solids volume
fraction f, is $r = r_0/f^{1/3}$.

The density of foamed material, ρ, depends on solids volume
fraction and density of solid material, ρ_0, as $\rho = f\rho_0$.

Substituting these values for foamed particle radius and density
into the standard expression for extinction by a sphere we find

$$\alpha = \frac{30}{4\rho_0 r_0 f^{2/3}}$$

At first glance one might conclude that reductions in the
initial unpuffed particle radius and solids fraction always result
in increases in extinction coefficient. This is not the case because
of the dependence that the extinction efficiency has upon initial
radius and solids fraction. In the Rayleigh-Gans limit, the extinction efficiency for a dielectric is proportional to the square
of the radius, and consequently the extinction coefficient is proportional to the radius, while in the large sphere limit the extinction coefficient is inversely proportional to radius. Somewhere
between these two size limits lies the maximum extinction. The
theory which encompasses this size and refractive index range is the
theory of anomalous diffraction (van de Hulst, 1957). Here for a
dielectric particle the extinction efficiency is a function of phase
shift defined as

$$\text{Phase Shift} = \frac{4\pi r}{\lambda}\ (m_e - 1)$$

where λ is the wavelength and m_e is the effective refractive index
for the foam. There are several effective medium theories that predict an effective refractive index for mixtures (see Granquist and
Hunderi, 1978), but here we simply assume a linear relationship
$m_e = 1 + f\ (m-1)$ where m is the refractive index of the solids. The
major peak in extinction efficiency in the anomalous diffraction
region has a value of about three and is located at a phase shift of
about four (van de Hulst, 1957). Substituting for the effective refractive index in the phase shift expression and setting the phase
shift equal to four, we get the following expression:

$$r_0 f^{2/3} = \frac{\lambda}{\pi\,(m-1)}$$

Inserting this along with the peak efficiency value of 3 into the expression for the extinction coefficient, we get a rough approximation for the maximum extinction by a non-absorbing foamed particle,

$$\alpha = \frac{9\pi(m-1)}{4\rho_0\lambda} \quad .$$

At a refractive index of 2, a density of 1.5 and a wavelength of 8 microns, the extinction for a puffed sphere becomes ~0.6 m^2 g^{-1}.

REFERENCES

Granquist, C. G. and Hunderi, O., 1978, *Phys. Rev. B*, *18*, 2897.
van de Hulst, H. C., 1957, "Light Scattering by Small Particles," John Wiley and Sons, New York.

SHAPE OF RAINDROPS

V. Ramaswamy and Petr Chýlek

State Univ. of N.Y. at Albany, Atmospheric Sciences
Research Center, Albany, NY 12222, and National
Center for Atmospheric Research, Boulder, CO 80307

ABSTRACT

Water drops found in natural rains acquire the simple, spherical forms at only the smallest sizes. Both laboratory and field research reveal that the larger sized raindrops undergo considerable deviation from a spherical shape. We review the evidence for deformed drops and show a technique for parameterizing the drop surface in accordance with the forces acting on falling raindrops.

1. EXPERIMENTAL EVIDENCE

All workers agree on the fact that raindrops are spherical at the smallest sizes. Photographs by Pruppacher and Beard (1970) of falling water drops reveal that drops of less than 140 μm in radius are spheres. They also noted that between radii of 140-500 μm, drops were measurably deformed into oblate spheroids. Magono's (1954) photograph shows a drop of 0.3 cm radius as being asymmetric about a horizontal plane through the center, being flattened at the bottom and rounded at the top. Koenig's (1965) photograph displays a pronounced concave depression in the base of a large drop (0.5 cm in radius). Radar returns also suggest the presence of non-spherical raindrops (Newell *et al.*, 1957; McCormick, 1968).

A convenient method of estimating the departure of a drop from a spherical form is to measure the ratio of the minor to the major axis (a/b). The "deformation ratio" so defined is examined in terms of an equivalent radius r_0, where r_0 is the radius of a sphere with a volume equal to that of the distorted drop. Figure 1 shows a typical deformed drop. The measurement of the deformation ratio at

small sizes is consistent with the results of various theoretical investigations. Figure 2 shows the variation of a/b at small r_0's.
The drops are spherical (a/b = 1) up to r $\simeq$ 100 µm and the perturbation from a spherical shape, *i.e.* the decrease in a/b, is only 1%
up to about 300 µm. In general, the deviation from a spherical shape
increases slowly with size, becoming faster from 250 µm onward.

Variations of the deformation ratio at large sizes is shown in
Figure 3. The mean values were estimated from various laboratory investigations (Best, 1947; Blanchard, 1950; Magono, 1954; Pruppacher
and Beard, 1970; Spilhaus, 1948; Kumai and Itagaki, 1954, as reported
by Jones, 1959). Also shown are *in situ* results from Jones (1959)
and a typical laboratory data obtained recently by Pruppacher and
Pitter (1971). The significant feature of all these results is the
progressive decrease of the deformation ratio with increasing drop
size, *i.e.* as r_0 increases, drops tend toward increasing oblateness.
Divergences at large sizes are due to drop instabilities and technical uncertainties. We note here that the largest drops observed
in natural rains have a radius of $\simeq$0.3 cm (Blanchard and Spencer,
(1970).

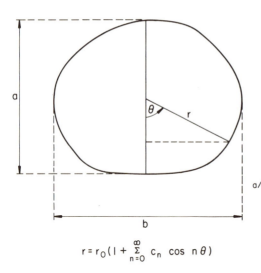

$$r = r_0 (1 + \sum_{n=0}^{\infty} c_n \cos n\theta)$$

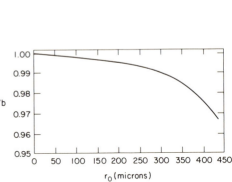

FIG. 1.   The shape of a large,
deformed raindrop (details ex-
aggerated) with minor axis a and
major axis b; r is the radius of
the deformed drop while r_0 is
the radius of the equivalent
(by volume) spherical drop.
The c_n's are first-order deformation coefficients.

FIG. 2.   The deformation ratio
a/b versus r_0 for small raindrop
sizes. The curve is the mean of
theoretical and experimental results, as reported by Pruppacher
and Beard (1970).

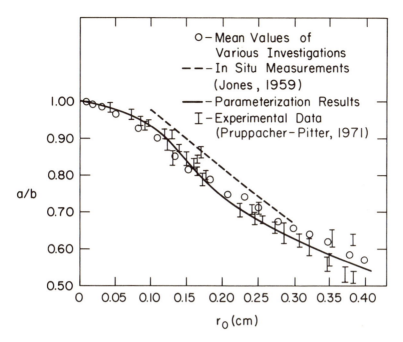

FIG. 3.   The deformation ratio a/b versus r_0 from various investigations for large raindrop sizes.

2.   DROP SURFACE PARAMETERIZATION

From observations of the deformation ratio, theoretical expressions have been derived by Pruppacher and Beard (1970) to predict the shape of raindrops. For small drops of radii between 140-500 μm, they give

$$\frac{a}{b} = \left[1 - \frac{9}{16} r_0 \, \rho_m \, V_\infty^2 / \gamma \right]^{1/2} \tag{1}$$

where ρ_m is density of water vapor saturated air in gm cm^{-3}, V_∞ is terminal velocity of the drop in cm s^{-1}, γ is the surface tension of water in erg cm^{-2}. For radii between 500-4500 μm, the following empirical expression is proposed (Pruppacher and Beard, 1970):

$$\frac{a}{b} = 1.030 - 0.124 \, r_0 \, , \tag{2}$$

where r_o is in mm. Both equations (1) and (2) are good fits to the
experimental data. However, the deformation ratio, though stating
the extent of deviation from a spherical shape, does not unambiguously
determine the exact shape at any particular size. For instance,
features such as flat bases and concave depressions are not expli-
citly revealed. Since our interest lies primarily in the scattering
of electromagnetic waves by raindrops, we need to know the exact form
of the surface of the scatterer. To achieve this, an analytic solution is required, based on the influence of the various physical
forces acting on a raindrop.

The important forces acting on a falling raindrop are external
aerodynamic drag, surface tension, hydrostatic pressure gradients
and internal circulation. Electrostatic influences are ignored,
following MacDonald (1954). A good review of these forces is given
by MacDonald and, hence, they will not be detailed here. For small
drops, the dominant force is surface tension, and this would consequently impart a spherical shape (Spilhaus, 1948). For large drops,
airflow patterns become quite complicated, and all the forces mentioned contribute significantly to the drop energy. Typically, this
significance is assumed at $r_o \simeq 0.05$ cm.

The problem of determining an analytic solution for the raindrop
shape is complicated by the fact that one of the major forces, the
aerodynamic pressure distribution over the surface of a falling drop,
is itself determined by the shape. As a first approximation, any
theory has to assume that the various forces have come into equilibrium and that the drops have attained their terminal velocities.
MacDonald assumes that the equilibrium shape of a large drop is
"that particular shape for which the joint action of the external
aerodynamic pressures and the surface pressure increments just produce an internal pressure distribution that satisfies the hydrostatic equation within the drop." Pruppacher and Pitter (1971) parameterized the drop surface in terms of an infinite series of first-
order perturbation terms *viz.*,

$$r = r_o(1 + \sum_{0}^{\infty} c_n \cos n\theta) \qquad (3)$$

where r is the radius of the deformed drop, r_o is the radius of the
equivalent spherical drop and c_n's are the deformation coefficients
(see Figure 1). The polar coordinate system is placed in a meri-
dional plane and is centered at the center-of-mass of the drop.
Representing all the forces in an infinite series of trigonometric
terms according to the above parameterization and substituting in
the pressure balance equation, it is possible to uniquely identify
the c_n's for each equivalent size. The infinite series in equation
(3) may be truncated at n = 9 without loss of precision. Values of

the deformation coefficients for certain specific sizes are tabulated
in Pruppacher and Pitter (1971) and Oguchi (1977). Each of the coefficients increases with size, resulting in increasing distortion
of the larger drops. Typical shapes of raindrops at four sizes, as
computed from equation (3), are shown in Figure 4.

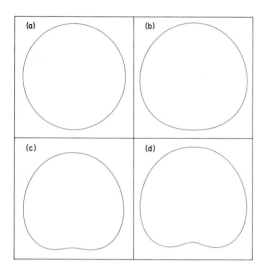

On the basis of this parameterization, drops with r_0 <
0.017 cm are spheres; drops with
r_0 = 0.017 - 0.05 cm are deformed
into oblate spheroids, the deformations increasing only slightly with size; drops with r_0 >
0.05 cm become asymmetric spheroids, with curvature decreasing
at the base and increasing at the
top; drops with $r_0 \simeq$ 0.2 cm
develop flat bases; and drops with
r_0 > 0.2 cm develop concave depressions in their bases.

Such a classification is consistent with the available photographic evidence. Figure 3 shows
comparison of the computed deformation ratios with various
measurements. The agreement is
quite good. Differences become
significant at r_0 > 0.3 cm which,
as noted earlier, is outside the
realm of natural raindrop sizes.

FIG. 4. Typical shapes of raindrops at various sizes as computed from equation (3): a)
spherical, r_0 = 0.017 cm; b)
asymmetric oblate, r_0 = 0.15 cm;
c) slight concave depression at
base, r_0 = 0.25 cm; d) pronounced concave depression at
base, r_0 = 0.30 cm.

3. DISCUSSION

This method of representing the drop surface, then, is a good
approximation for raindrops. The important assumption in the analysis
is that the forces acting on a raindrop have come to an equilibrium
and that the drop is falling at its terminal velocity. Such a con-
cept of an equilibrium shape is probably appropriate since raindrops
are known to reach terminal velocities in a very short time. In
natural situations, however, it is possible that the equilibrium
shapes would not be true of all drops of a particular equivalent size.
Blanchard (1950) observed drops in the laboratory undergoing periodic
transitions between various spheroidal forms. Jones (1959) mentions
the possibilities of rotational and vibrational motions of drops
about their figure axes. Best (1947) suggests oscillations about a

mean shape. In such cases, oblate forms would coexist with other
spheroidal forms.

The question of the relative occurrence of various shapes in
natural rains is resolved to some extent by the *in situ* photographs
of Jones (1959). In a study of drops with r_0 > 0.1 cm observed during
a rainstorm, the predominant shapes were oblate spheroidal forms,
occurring as pure oblate spheroids or deformed oblate spheroids.
Other forms were prolates or irregular ellipsoids. In Figure 3 the
average deformation ratios of all the samples is shown. The prime
feature is the decrease in deformation ratio with increasing size,
implying the overwhelming presence of oblate forms. The fact that
the average deformation ratio is always $\leq$ 1 for all sizes indicates
that oblate forms may be regarded as the average raindrop shapes
under ambient conditions. Prolate forms may be anticipated as a
secondary feature and, consequently, in lesser proportions. We have
already seen that the analytical results, too, predict the oblate
form (a/b < 1) as the equilibrium shape. Thus, the average shape
may be equated with the equilibrium shape. Such an approximation,
in which all drops of all sizes are regarded as having oblate forms,
would be particularly useful when considering drop-size distributions.

In addressing the problem of electromagnetic scattering mechanisms and remote sensing measurements, it must be pointed out that
in our treatment we have the raindrops oriented parallel or perpendicular to an arbitrary surface level. In real physical situations,
local wind fields may force the drops to be "canted" at arbitrary
angles with respect to the ground. In conclusion, the electromagnetic
wave propagation in rain involves interaction with non-spherical deformed oblate drops at the larger sizes. Current treatments of the
raindrops in radar meteorology and communication sciences regard
drops of all sizes usually as spheres or, at best, as symmetric
oblates. Since a typical raindrop size distribution contains drops
$\geq$ 0.1 cm in equivalent radius, it is necessary to reinterpret the
scattering characteristics in terms of the nonsphericity of the drops.

Acknowledgement. This research was supported in part by the
Atmospheric Research Section of the National Science Foundation.

REFERENCES

Best, A. C., 1947, Air Ministry, Great Britain Meteor. Res. Comm.,
    M. R. P. 330.
Blanchard, D. C., 1950, *Trans. Amer. Geophys. Union*, *31*, 836.
Blanchard, D. C., 1970, *J. Atmos. Sci.*, *27*, 101.
Jones, D. M. A., 1959, *J. Meteor.*, *16*, 504.
Koenig, R., 1965, *J. Atmos. Sci.*, *22*, 448.
MacDonald, J. E., 1954, *J. Meteor.*, *11*, 478.

Magono, C., 1954, *J. Meteor.*, <u>11</u>, 77.

McCormick, G. C., 1968, *in* "Proc. 13th Conf. Radar Meteor.," Boston, AMS, 340.

Newell, R. E., Geotis, S. G., and Fleisher, A., 1957, MIT Dept. Meteor., Res. Rept. No. 28, p. 57.

Oguchi, T., 1977, *Radio Sci.*, <u>12</u>, 41.

Pruppacher, H. R. and Beard, K. V., 1970, *Quart. J. Roy. Meteor. Soc.*, <u>96</u>, 247.

Pruppacher, H. R. and Pitter, R. L., 1971, *J. Atmos. Sci.*, <u>28</u>, 86.

Spilhaus, A. F., 1948, *J. Meteor.*, <u>5</u>, 108.

ATMOSPHERIC ICE CRYSTALS

Bernard Vonnegut

State Univ. of N.Y. at Albany, Atmospheric Sciences Research Center

Albany, New York 12222

ABSTRACT

Classifications of naturally occurring atmospheric ice crystals are presented, and their formation and occurrence are discussed.

More than one-half the mass of the earth's atmosphere has a temperature below 0°C. Ordinary hexagonal ice is the stable phase of water under these conditions of pressure and temperature. It is, therefore, a common constituent of the atmosphere. Even during the hottest weather, the 0° isotherm is never more than about 5 km above the earth's surface.

Though ice is the stable phase, unstable supercooled water droplets often are present down to temperatures as low as -40°C. Laboratory and field experiments indicate that this temperature, the so-called "Schaefer point", is the limit to which even the purest water can be supercooled. Below this temperature water can exist only in the form of a solid or a gas. In the Polar regions and in the upper troposphere, all condensed water phase is necessarily ice, for liquid water cannot exist at these low temperatures. The lowest temperatures in the atmosphere are to be found in the Polar regions in the winter and in the upper portions of giant thunderstorms where temperatures less than 200°K have been observed.

Atmospheric ice crystals occur in a very wide variety, the shapes and sizes determined by two factors, the initiation or nucleation of the ice crystal and its subsequent growth.

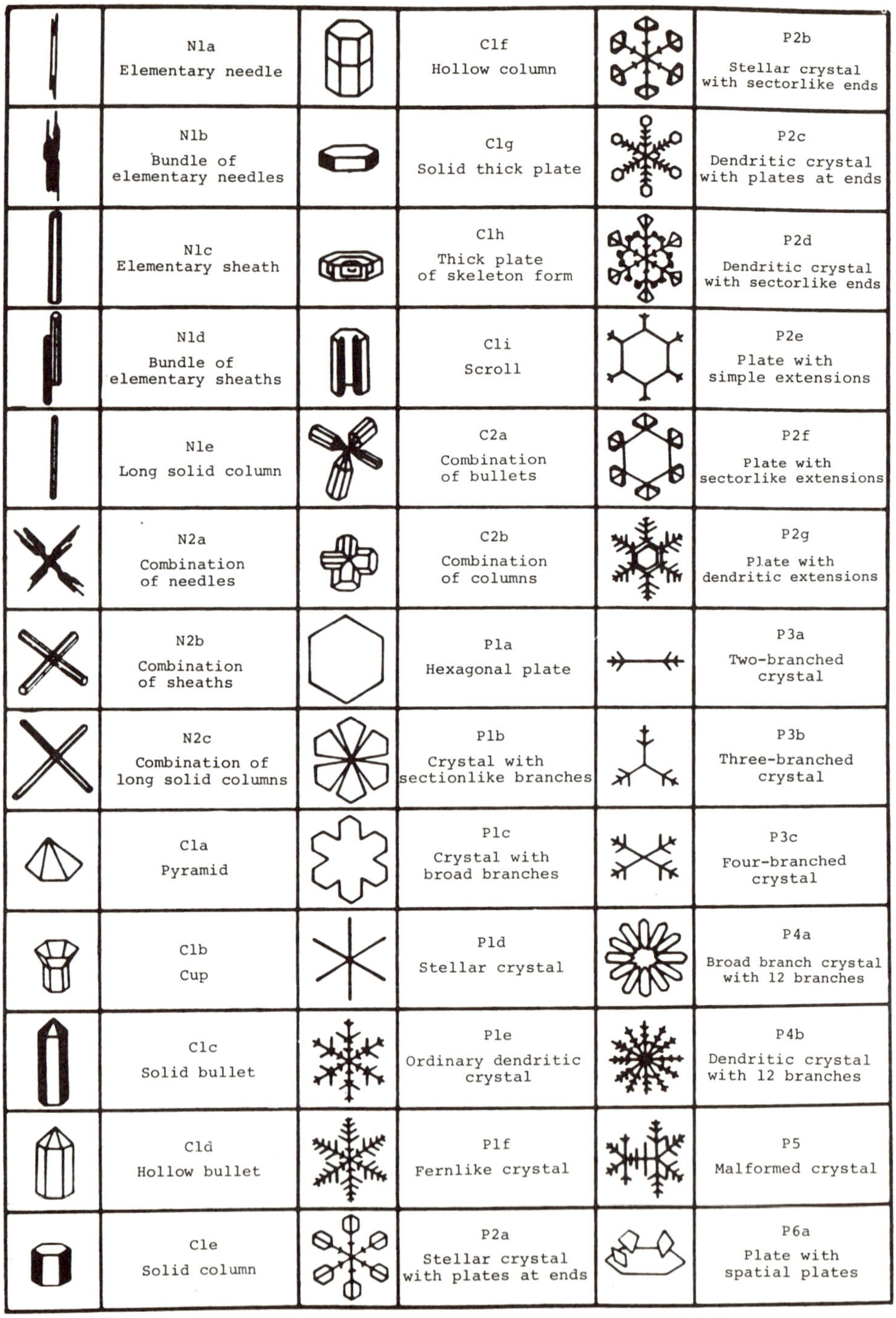

	N1a Elementary needle		C1f Hollow column		P2b Stellar crystal with sectorlike ends
	N1b Bundle of elementary needles		C1g Solid thick plate		P2c Dendritic crystal with plates at ends
	N1c Elementary sheath		C1h Thick plate of skeleton form		P2d Dendritic crystal with sectorlike ends
	N1d Bundle of elementary sheaths		C1i Scroll		P2e Plate with simple extensions
	N1e Long solid column		C2a Combination of bullets		P2f Plate with sectorlike extensions
	N2a Combination of needles		C2b Combination of columns		P2g Plate with dendritic extensions
	N2b Combination of sheaths		P1a Hexagonal plate		P3a Two-branched crystal
	N2c Combination of long solid columns		P1b Crystal with sectionlike branches		P3b Three-branched crystal
	C1a Pyramid		P1c Crystal with broad branches		P3c Four-branched crystal
	C1b Cup		P1d Stellar crystal		P4a Broad branch crystal with 12 branches
	C1c Solid bullet		P1e Ordinary dendritic crystal		P4b Dendritic crystal with 12 branches
	C1d Hollow bullet		P1f Fernlike crystal		P5 Malformed crystal
	C1e Solid column		P2a Stellar crystal with plates at ends		P6a Plate with spatial plates

FIG. 1.   Classification of ice crystals by Magono and Lee (1966).

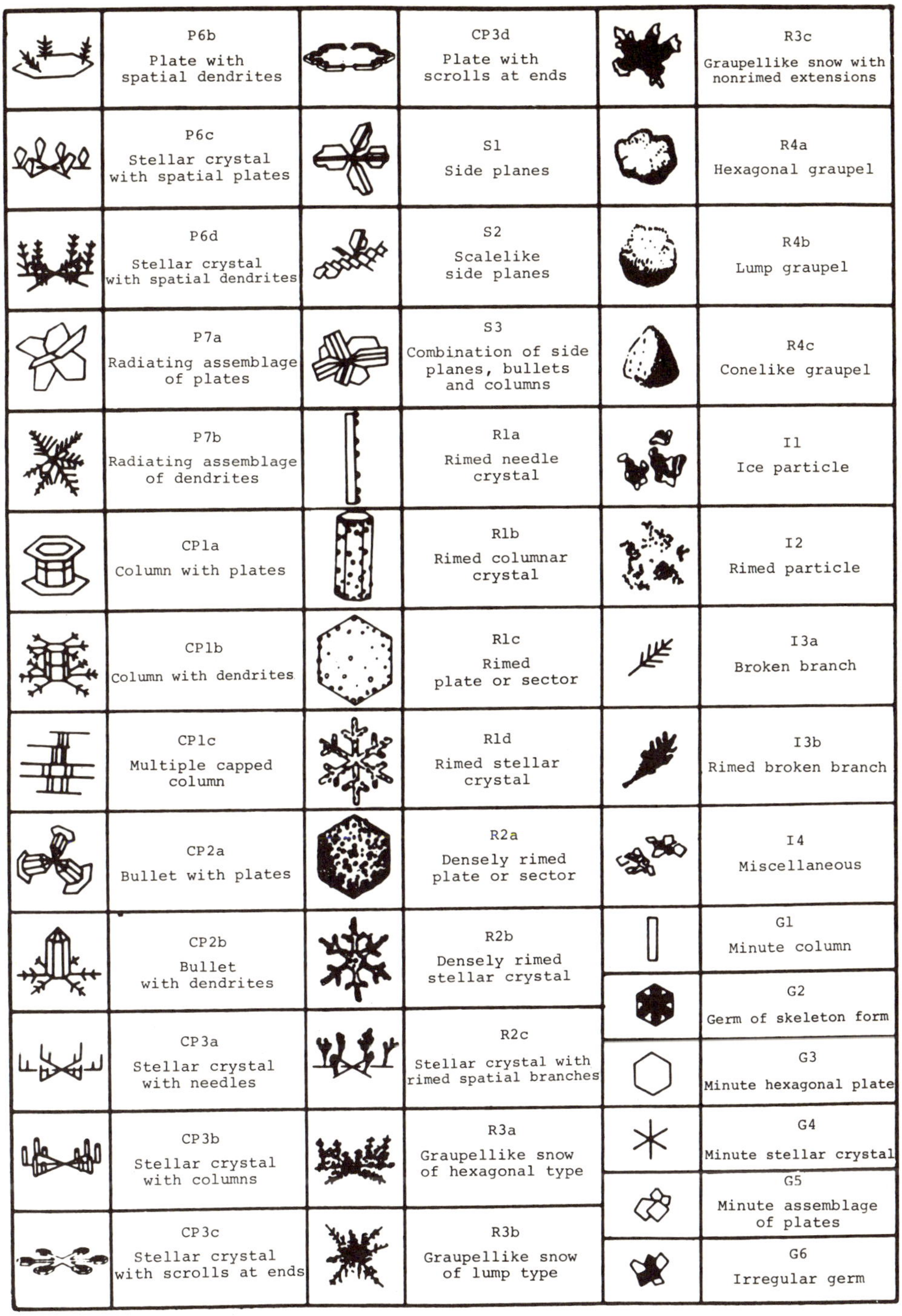

	P6b Plate with spatial dendrites		CP3d Plate with scrolls at ends		R3c Graupellike snow with nonrimed extensions
	P6c Stellar crystal with spatial plates		S1 Side planes		R4a Hexagonal graupel
	P6d Stellar crystal with spatial dendrites		S2 Scalelike side planes		R4b Lump graupel
	P7a Radiating assemblage of plates		S3 Combination of side planes, bullets and columns		R4c Conelike graupel
	P7b Radiating assemblage of dendrites		R1a Rimed needle crystal		I1 Ice particle
	CP1a Column with plates		R1b Rimed columnar crystal		I2 Rimed particle
	CP1b Column with dendrites		R1c Rimed plate or sector		I3a Broken branch
	CP1c Multiple capped column		R1d Rimed stellar crystal		I3b Rimed broken branch
	CP2a Bullet with plates		R2a Densely rimed plate or sector		I4 Miscellaneous
	CP2b Bullet with dendrites		R2b Densely rimed stellar crystal		G1 Minute column
					G2 Germ of skeleton form
	CP3a Stellar crystal with needles		R2c Stellar crystal with rimed spatial branches		G3 Minute hexagonal plate
	CP3b Stellar crystal with columns		R3a Graupellike snow of hexagonal type		G4 Minute stellar crystal
					G5 Minute assemblage of plates
	CP3c Stellar crystal with scrolls at ends		R3b Graupellike snow of lump type		G6 Irregular germ

FIG. 2. Classification of ice crystals by Magono and Lee (1966).

The purest supercooled water spontaneously freezes at temperatures above −40°C, even in the complete absence of foreign particulate matter. Similarly, it is observed that at low temperatures and
high supersaturations gaseous water vapor can condense to form ice
crystals in the absence of particulate matter. This kind of ice
crystal formation that does not involve the presence of any foreign
substances is known as homogeneous nucleation and can take place in
the colder portions of the atmosphere.

More commonly, ice crystal formation both in supercooled water
and in air supersaturated with respect to ice occurs on small particles, known as ice-forming nuclei. These particles, which catalyze
freezing, vary greatly in their efficacy. Small silver iodide particles and certain natural clays or organic particulate matter are
capable of serving as centers for ice crystal formation at temperatures of less than 1°C below the melting point of ice. Other
particles serving as nuclei may require temperatures as low as −20°C
or −30°C before they are capable of initiating ice crystal formation.

The ultimate shape of a supercooled drop that freezes is strongly
dependent on the location of the nucleus where freezing begins. For
example, when freezing begins at the drop surface, a solid shell of
ice can form that will then rupture when the interior of the drop
freezes and expands. If nucleation occurs in the interior of the
drop, freezing may take place without the build-up of internal pressures and subsequent fracture.

When ice crystals form in clear air supersaturated with respect
to ice on ice nucleating aerosol particles, the nucleation process
can have an important effect on the ultimate shape of the ice particle by interfering with growth in some direction or causing stresses
and imperfections in the crystalline lattice.

Once an ice crystal has been started either by homogeneous or
heterogeneous nucleation, its shape is determined by the rate at
which it grows. Its growth can occur by crystal growth arising from
the diffusion of water molecules from supersaturated vapor, by the
accretion of supercooled water droplets and subsequent crystalli-
zation, and by the aggregation of crystals resulting from surface
adhesion or electrical forces. The rate of growth of ice crystals
is strongly dependent on the temperature and partial pressure of
water vapor. These variables, therefore, have a large influence on
the ultimate shape of the crystal. Laboratory experiments have established that trace materials in the atmosphere in concentrations
as low as a few parts per million can have a significant effect on
the relative rate of growth of the crystal in different directions.
The speed at which the crystal is falling through the atmosphere obviously has a large effect on the rate of diffusion of water vapor
to the surface of the growing crystal and on the rate at which the
falling crystal may collide with solid or liquid cloud particles.

Laboratory experiments show that strong electric fields, of the sort
that occur within thunderstorms, may have a pronounced effect on ice
crystal growth rates and, under some conditions, can cause the formation of long crystalline fibers.

When crystal growth is occurring within a supercooled liquid
drop, high supersaturations of dissolved gas can occur at the crystal-
liquid interface. The resultant bubbles that sometimes form can play
a large part in the growth and optical properties of the resultant
frozen particles.

In the case that an ice crystal that has formed in the atmosphere enters a region unsaturated with respect to ice, evaporation
will occur that can modify its shape. In some cases dendritic structures may become so weakened that the crystal can break into smaller
pieces. In the case where an ice crystal moves into air above 0°C,
melting will take place that can profoundly modify the shape and
optical properties of the crystal. When snowflakes fall below the
0°C isotherm, the liquid water layer produced on their surfaces by
melting significantly increases the scattering of microwave radiation,
causing the formation of the so-called radar "brightband" effect at
the freezing level.

The diversity of the various forms of frozen particles occurring
in the atmosphere is illustrated by Figures 1 and 2, showing the
classification of ice crystals by Magono and Lee (1966).

Comment (MCKECHNEY): The Air Force Geophysics Laboratory Cloud
Physics Aircraft has measured a large population of non-pristine ice
crystals in clouds. (D. J. Varley and A. A. Barnes, Jr., "Cirrus
Particle Distribution Study, Part 4," AFGL TR 79-0134, June, 1979,
and references therein). There is a geographical bias. Over the
Pacific, where one would expect fewer aerosols, more simple pristine
crystals were found. However, over the continental U.S. the large
number of aerosols leads to crystal masses or conglomerates. Even
in cirrus, where there is an optical phenomena suggestive of simple
crystals, the larger conglomerates were found.

REFERENCE

Magono, C., and Lee, C. W., 1966, Journal of the Faculty of Science,
    Hokkaido University, Japan, Series VII, Volume II, No. 4, p. 324-325.

PHYSICAL PROPERTIES OF ATMOSPHERIC PARTICULATES

Roger J. Cheng

State Univ. of N.Y. at Albany, Atmospheric Sciences Research Center

Albany, New York   12222

Microscopic investigations of atmospheric particulates reveal a complex variation with respect to size and shape. Emission from combustion processes contributes more than 70% of the fine particle burden in the atmosphere. Flyash from coal-fired boilers have a relatively smooth surface in contrast to flyash from oil-fired boilers which have a rough, honeycomb-like surface. Particles from automobile exhaust and smoke are extremely small, spherical, and fairly uniform in size. In addition they always exhibit a chain-like aggregation.

Classification of fine structure of ice crystals is based on environmental conditions such as temperature and degree of super-saturation. However, recent evidences have shown that it is also dependent on the origin of their freezing nuclei. Selected micrographs and classification, based on major sources, size, and major chemical components, from the "ASRC Atmospheric Particulates Atlas", are presented with self explanatory captions.

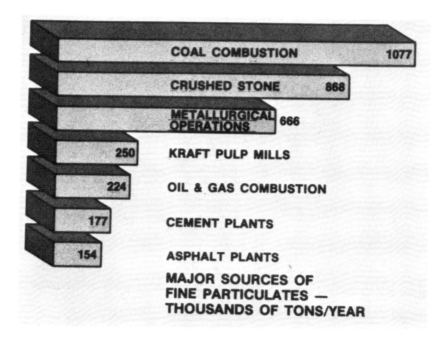

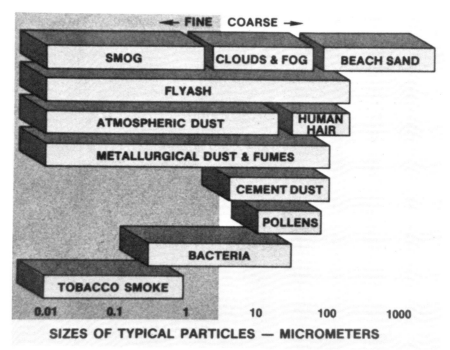

FIG. 1.   Major sources and sizes of atmospheric particulates.

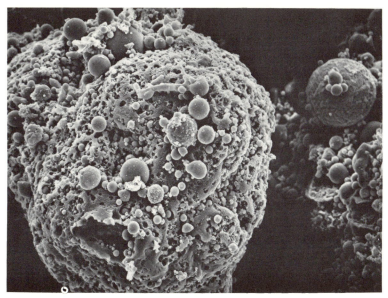

FIG. 2. Flyash from a coal-fired power plant. Size range: 0.01 μm–100 μm. Microscopy: SEM. Major chemical components: Si, S, K, Ca, Ti, Fe. Physical properties: Fairly smooth sphere.

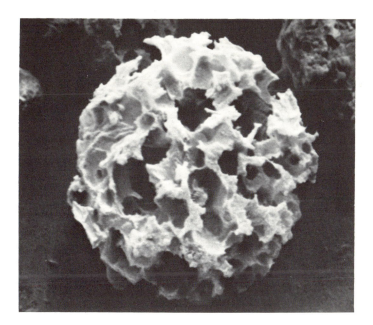

FIG. 3. Flyash from an oil-fired power plant. Size range: 0.01 μm–100 μm. Microscopy: SEM. Major chemical components: Si, S, Ca, V, Fe. Physical properties: Very rough honeycomb-like sphere.

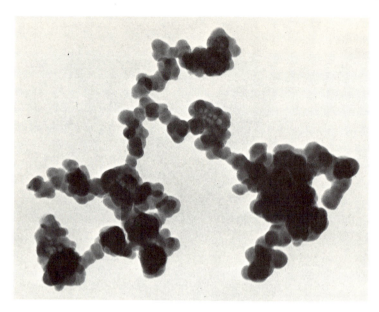

FIG. 4.   Emission from automobile exhaust. Size range: 200Å–500Å.
Microscopy: TEM. Major chemical components: Cl, Br, Fe, Pb. Physical
properties: Spherical particle with chain-like aggregation.

FIG. 5.   Sand dust. Size range: 0.01 μm–100 μm. Microscopy:
Optical. Major chemical components: Al, Si, Ca, Fe. Physical
properties: Sharply angular chips.

FIG. 6.   Particles from cement plant.   Size range: 5 μm–100 μm.
Microscopy: Optical.   Major chemical components: Al, Si, K, Ca, Fe.
Physical properties: Rounded grains with various density.

FIG. 7.   Particles from steel plant.   Size range: 0.0 μm–100 μm.
Microscopy: Optical.   Major chemical components: Al, Si, Ca, Fe, Zu
K.   Physical properties: Angular chips with spheres.

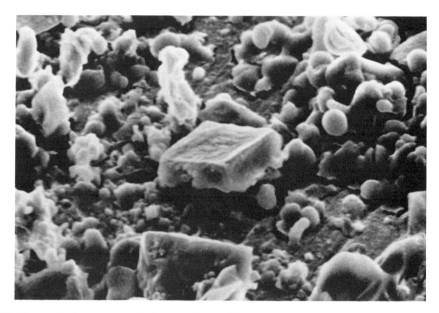

FIG. 8.   Salt crystal from mist of sea water.   Size range: 0.1 µm-
100 µm.   Microscopy: SEM. Major chemical components: NaCl.   Physical
properties: Cubic crystal.

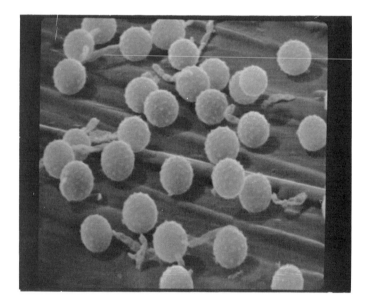

FIG. 9.   Pollen – ragweed.   Size range: 10 µm–100 µm.   Microscopy:
SEM.   Physical properties: Spherical with surface covered with
arrays.

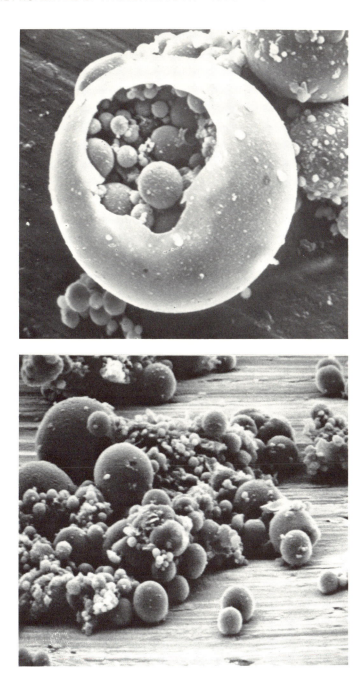

FIG. 10.   Flyash from a coal-fired power plant.   Size range: 0.01 μm–
100 μm. Microscopy: SEM. Major chemical components: Si, S, K, Ca,
Ti, K. Physical properties: Fairly smooth spheres.

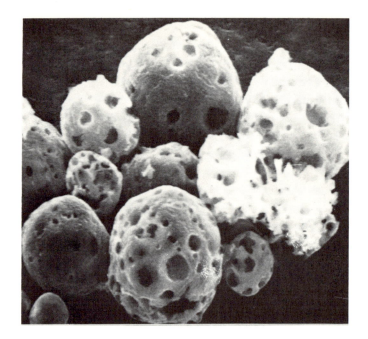

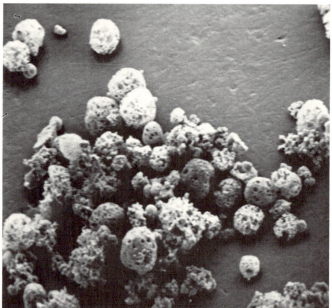

FIG. 11.   Flyash from an oil-fired power plant.   Size range: 0.01 μm–
100 μm. Microscopy: SEM. Major chemical components: Si, S, Cg, V,
Fe. Physical properties: Very rough honeycomb-like spheres.

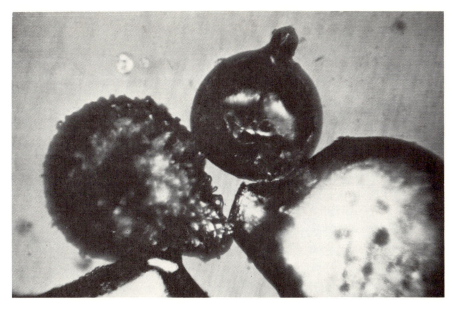

FIG. 12.   Frozen water droplets.   Size range: 20 µm-1 µm.   Microscopy: Optical. Physical properties: Spheres with spicules and cloud
droplets on surface.

FIG. 13.   Particulates from an American home.   Size range: 1 µm-
100 µm. Microscopy: Optical. Major chemical components: Cl, Si, Ca,
Fe, K, S, Al. Physical properties: Fine particles with various fibers
and dandruff.

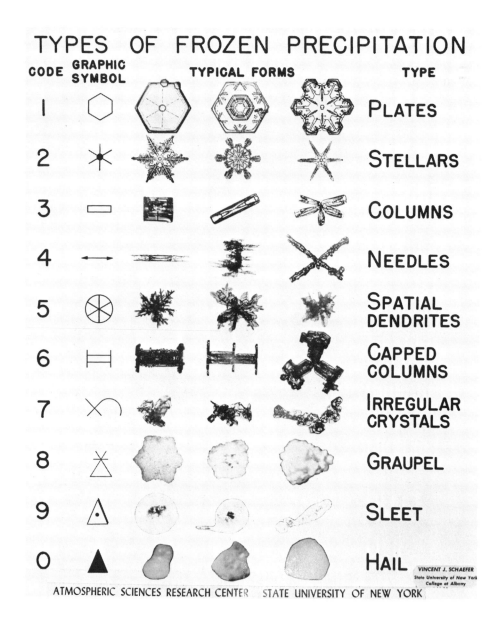

FIG. 14.   Frozen precipitation – ice crystals.   Size range: A few μm–
300 μm.   Microscopy: Optical.

SOME CHARACTERISTICS OF THE ANTARCTIC AEROSOL

A. Hogan, A. M. Pitaniello, R. Cheng, S. Barnard, W. Winters

State Univ. of N. Y., Atmospheric Sciences Research Center

Albany, New York   12222

J. A. Samson

State Univ. of N. Y., Dept. of Atmospheric Sciences

Albany, New York   12222

ABSTRACT

Aerosol concentrations have been systematically measured at the South Pole for five years. These climatological measurements have been supported by vertical and horizontal profiles obtained in summer with aircraft flights. Aerosol size data has been obtained with diffusion batteries, electrostatic precipitators, and cascade impactors.

There is a strong (5 to 1) seasonal variation in surface aerosol concentrations, with the maximum aerosol concentration generally occurring with the beginning of summer mixing in November. Vertical profiles consistently show the greatest aerosol concentrations to occur in the moist layer, a few hundred meters above the surface, and then diminishing quite steadily with altitude.

Examination of collected particles by light and electron micro-scopy shows them to be soluble, with refractive index of ~1.54 and often with the appearance of flattened drops. The maximum particle radius found was 3 μm and the peak volume concentration occurred at 2 μm radius. The size distributions are of similar slope to those measured over the Weddell Sea by Meszaros and in Tasmania by Bigg. We interpret this as evidence that the Southern Ocean and especially

the Weddell Sea as the source of these particles. The particles
arriving at the station are then quite probably the residue of
evaporated clouds, and the aerosol is representative of that
occurring at the end point of many of nature's particle removal
mechanisms.

1.   INTRODUCTION

A systematic series of surface aerosol observations were begun
at the South Pole in 1974, through cooperation of D. Pack of the
NOAA-GMCC program. Aerosol measurements, including some size
measurements, have been made twice daily since that time using a
Pollak (1959) counter as an aerosol sensor; and the diffuser denuder
technique of Rich (1966), the diffusion technique of Sinclair (1972),
and the filter technique of Spurny *et al.* (1969) to determine size.
Light scattering and impaction techniques have been used in summer
months to determine some of the properties of the particulate fraction
of radius greater than .25 μm. These measurements are made concurrent
with meteorological observations and soundings, and some transport
mechanisms have been hypothesized.

Analysis of initial data indicated that a relatively moist layer
is consistently present above the station and that this is the most
favorable layer for transport of particulate material from the peri-
phery to the interior of Antarctica. Preliminary aircraft investigations did indicate that aerosol concentrations were greater in this
layer than those found at the surface at the South Pole.

2.   RESULTS OF EXPERIMENTS

2.1 Vertical and Horizontal Profiles
of Aerosol Concentrations over Antarctica

Hoffman and Rosen have obtained several vertical profiles of
total and light scattering aerosol over McMurdo Station (78S) and the
South Pole. The clear weather required for balloon launches limits
their information to high pressure systems over McMurdo, but is quite
representative of most summer days over the South Pole. They have
found small aerosol concentrations near the ice surface, which increase in the humid layer on most occasions, and have found enriched
aerosol layers aloft on several occasions. We have obtained several
vertical profiles over the Ross Sea, the Ross Ice Shelf at the South
Pole, and other places over the ice in the interior of Antarctica;
and have transected from open water to the pole at several altitudes.

Typical vertical profiles of summer variation in temperature
humidity and aerosol concentration over the South Pole are shown in
Figure 1. From these and other soundings we have concluded that the

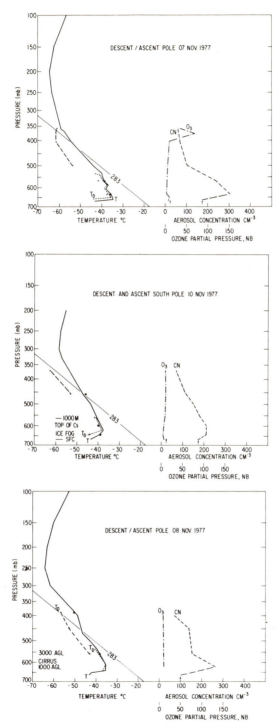

FIG. 1.   Typical vertical profiles of temperature, dew point, aerosol concentration and ozone over the South Pole

most favorable levels for transport of moisture and particles to the polar plateau are in the lower and mid troposphere. We found no evidence for stratospheric transport in a limited number of flights.

2.2 Meteorological Variation in Aerosol Concentration

It became very apparent during the first year of aerosol measurements in Antarctica that a strong seasonal variation in aerosol concentration is present at the surface. Mean summer concentrations exceed mean winter concentrations by a fivefold factor, and the maximum winter values are less than mean summer values.

Five years of observational data were stratified by wind direction at the time of observation and used to construct Figures 2 and 3. Analysis of these figures indicates that when a high pressure system is centered over North East Antarctica (the most usual winter condition) a weak gradient plus catabatic wind arrives at the station,

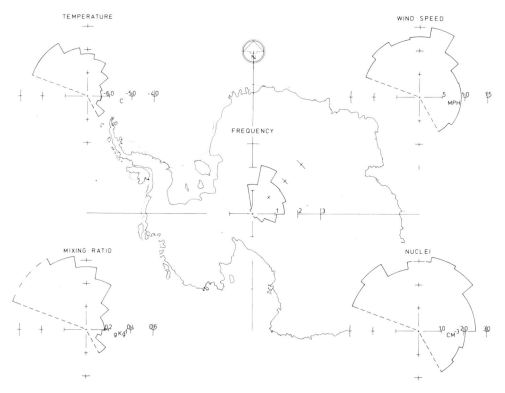

FIG. 2.   Climatology of transport of heat, water vapor and particles to the South Pole in winter.

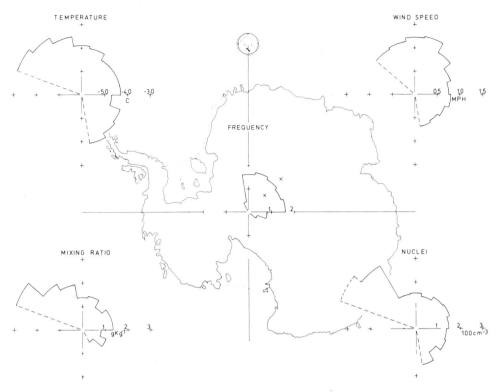

FIG. 3.   Climatology of transport of heat, water vapor and particles
to the South Pole in summer.

along 030° longitude. If the high moves to the vicinity of the
Greenwich Meridian, the surface winds are along 330° which is a
short path from the Weddell Sea, and warm, wet, aerosol laden air
arrives at the South Pole. When the high center is along the date
line the air has a long trajectory over ice and arrives at the
station along 090°, depleted of heat, water vapor and particles.

2.3   Aerosol Size and Physical Properties

A size analysis of aerosol which arrived with winds from 0° to
090° during the period 10-13 December 1978 is shown in Figure 4.   The
average aerosol concentration during this period was $106/cm^3$ of which
$44/cm^3$ were charged. Particles of .25 µm or larger were captured
inertially with a May (Casella) cascade impactor and analyzed micro-
scopically. All particles were extremely hygroscopic, and immediately
grew to solution drops when exposed to humid air. Application of
index of refraction oils in the range n = 1.52-1.56 caused almost all

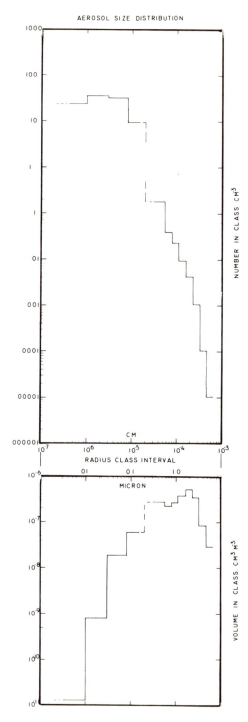

FIG. 4.   The volume and number distributions of aerosol particles
at the surface, South Pole, December 10-13, 1978.

particles to become invisible. Application of oils of n < 1.50 or
n > 1.60 caused no change in appearance. About 2% of all particles
examined appeared to be opaque or exhibit some birefringence. After
completion of light microscopy, a portion of several slides were gold
coated and examined with the scanning electron microscope. A typical
field is shown in Figure 5 abcd. At high magnification, and when the
substrate is tilted, the particles have a flattened appearance similar
to the residue of solution droplets.

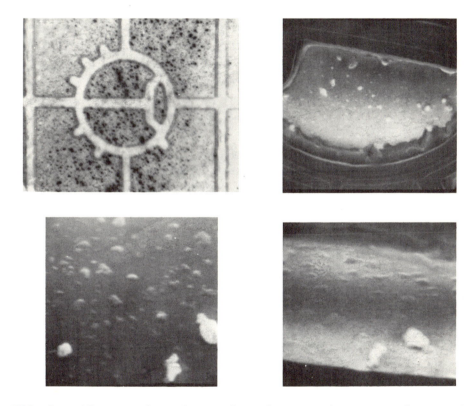

FIG. 5. Microscopic and scanning electron microscope photographs
of particles collected at the South Pole. Most seem to have the
appearance of solution drops.

3.   CONCLUSIONS

Analysis of the meteorological processes which may cause transport of particles to the South Pole, and the physical properties of the particles causes us to conclude that a majority of these particles are solution drops which are the residue of evaporated clouds. Comparison of the size distribution obtained at Pole with those obtained over the southern ocean by Bigg and Meszaros indicates that the South Ocean, and from trajectory analyses most likely the Weddell Sea is the source of these particles. We hypothesize that the aerosol observed at the South Pole has physical properties that are changing slowly with time - that is the aerosol has reached an "end point" after suffering many meteorological (especially cloud) removal processes. We also hypothesize that aerosol in the mid latitude troposphere might reach a similar "end point" while suffering exposure to similar cloud processes while traveling vertically in the atmosphere.

REFERENCES

Pollak, L. W., 1959, *Geof. Pura e Appl.* *43*, 285-301.

Rich, T. A., 1966, *J. Rech. Atm.* *2*, 79-86.

Sinclair, D., 1972, *J. Amer. Hygiene Assoc.* *33*, 729.

Spurny, K., Lodge, Jr., J. P., Frank, E. R., and Sheesly, D. C., 1969, *Envir. Sci. Tech.* *3*, 464-468.

EXAMPLES OF REALISTIC AEROSOL PARTICLES

COLLECTED IN A CASCADE IMPACTOR

David C. Woods

NASA Langley Research Center

Hampton, Virginia   23665

ABSTRACT

In this paper some scanning electron microscope photomicrographs showing examples of a variety of particulate aerosol shapes are presented. Compositions of particles also are presented. These particles represent samples from a number of different sources and locations including: solid propellant rocket motor plumes, active volcano plumes, the lower stratosphere over Sondrestrom, Greenland, and the upper troposphere over northern Texas. The particles were collected from aboard an aircraft with a cascade impactor which classified them according to aerodynamic size into 10 size intervals ranging from submicron to greater than 25 micrometers in diameter. The cascade impactor also served to measure the mass concentration as a function of particle size. The variety of shapes and compositions found among these particles suggest difficulties in obtaining reliable size distribution data from light scattering measurements.

1. INTRODUCTION

The scattering of light by small particles suspended in the atmosphere provides information on the size distribution, the concentration, and in some instances the composition of the particles. Light scattering techniques are often used to measure these parameters both *in situ* and remotely. In many cases the instrumentation employed in the measurements is calibrated using spherical particles as standards, and in other cases the measurements are related to theory in which it is assumed that the sample particles are spherical. It is, therefore, very convenient to treat atmospheric aerosols as if they are spherical in shape when employing scattering techniques.

It is generally appreciated, however, that the scattered intensity in
a given direction by a single non-spherical particle depends not only
on the particle's shape, but on its orientation as well. Thus, ignoring the fact that many aerosols are non-spherical could lead to
serious errors in the interpretation of data from light scattering
measurements.

    In this paper some examples of aerosol samples are presented
which illustrate the various shapes of particles found in the atmosphere. These samples were collected from several different sources
and locations including: rocket exhaust plumes, active volcano
plumes, the lower stratosphere over Sondrestrom, Greenland, and the
upper troposphere over northern Texas.

    The particles were collected *in situ*, from aboard an aircraft,
with a quartz crystal microbalance (QCM) cascade impactor (Chaun,
1976). The cascade separates the particles into 10 aerodynamic size
ranges and collects them on the surface of piezoelectric crystals
which produce real-time electrical signals corresponding to the mass
concentrations of the particles. Thus, a measurement of particle mass
concentration as a function of aerodynamic size is obtained. Table
1 gives the 50 percent efficiency points for the cascade impaction
stages for spherical particles with a mass density of 2 gm cm^{-3}. In
addition to size distribution data, information on the particle's
shapes and elemental composition was obtained by postflight analysis.
The crystals on which the particles were deposited were subjected to

TABLE 1

50 PERCENT EFFICIENCY POINTS FOR THE
STAGES OF THE QCM CASCADE IMPACTOR

Stage Number	50% Efficiency Point[*] (μm)
1	25
2	12.5
3	6.4
4	3.2
5	1.6
6	0.8
7	0.4
8	0.2
9	0.1
10	0.05

[*]The particle diameter for which the impaction efficiency is 50 percent for spherical particles with mass density of 2 gm per cubic
centimeter.

scanning electron microscope (SEM) analysis. The SEM was equipped
with an energy dispersive x-ray attachment for identifying elements
with atomic numbers greater than 10. The photomicrographs which show
the morphology of these particles are presented herein.

2.   LOWER STRATOSPHERE

In November, 1978 the QCM cascade impactor was flown over an
area near Sondrestrom, Greenland, aboard NCAR's Sabreliner jet along
with other sensors as part of a coordinated effort to obtain aerosol
data in support of NASA's SAM II (Stratospheric Aerosol Measurement)
ground truth measurements. Data were collected at several altitudes
ranging from 11.3 km to 12.5 km. The SEM photomicrographs in Figures
1A-1D show some of the particles collected in stage 7 of the cascade
impactor. (A) shows a low magnification of a 3.2×10^4 square micrometer area. (B) shows a higher magnification of the same area giving
more detail. Most of the particles appear to be irregular shaped
agglomerates which probably have very low mass densities to have
reached stage 7 in the cascade impactor. Some of these particles are
much larger than 5 μm in diameter. Stage 7 has a theoretical 50 percent cut point of 0.4 μm. The few very small singlets seen in (B),
(C), and (D) are about the correct size for stage 7. The large particles in (B), (C), and (D) clearly illustrate the presence of non-
spherical particles in the lower stratosphere, at least at this latitude. Although they may be less abundant than the more typical
spherical shaped H_2SO_4 droplets expected to be present in the stratosphere (Junge, 1963), they can contribute to the signals measured in
light scattering experiments and could, therefore, be important in
interpretation of data. The particles identified as A and B in (B)
show only traces of Al and S in their x-ray energy spectra. Both the
Al and S peaks are quite weak considering the size of the particles
and since no other peaks are present in the spectra, the particles
probably consist of some elements with atomic numbers less than 11
which the SEM cannot identify. It may be possible that these are
some form of hydrocarbons produced by aircraft.

3.   TROPOSPHERE

The samples shown in Figures 2A and 2B were collected over an
area in Texas near Dallas during a SAGE (Stratospheric Aerosol and
Gas Experiment) ground truth underflight. Each photograph represents
samples taken on stage 6 of the cascade impactor at both 4.9 km and
7.6 km, (A) near sunset on March 11, 1979, and (B) on March 13, 1979,
near sunrise. The particles in (A) consist of large irregular shaped
agglomerates which show large amounts of iron with a trace of sulfur
in the x-ray energy spectra. The small particles are about the right
size for this stage and are more nearly spherical in shape but are
not really smooth spheres. They show the same elements in the x-ray

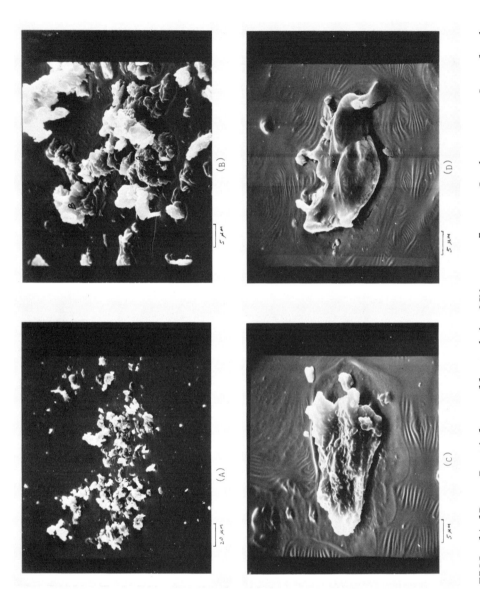

FIGS. 1A–1D.   Particles collected in QCM stage 7 over Sondrestrom, Greenland.

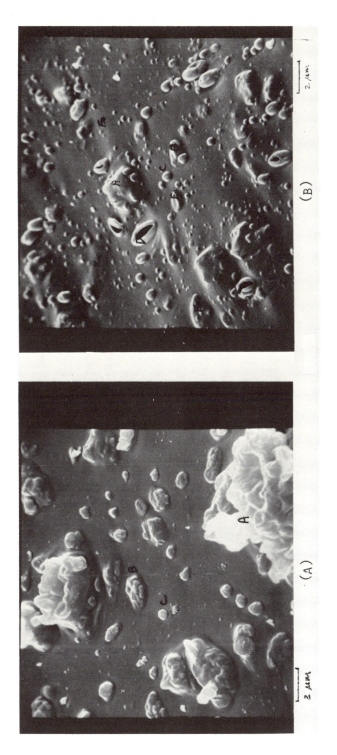

FIGS. 2A and 2B.   Particle collected in QCM stage 6 over northern Texas.   (A) March 11, 1979,
(B) March 13, 1979.

spectra as the large ones. (B) shows particles collected in stage 6
on March 13. These particles are quite different in morphology.
There are fewer of the large agglomerates present. There is a large
number of small spheres and quite a few nearly round hollow balls
that appear to have burst or collapsed, probably from the SEM vacuum.
The agglomerates such as particles A and B show calcium, magnesium,
aluminum, and iron where these hollow particles show the same spectrum
as the background indicating no identifiable elements.

4. AUGUST '77--TITAN

NASA Langley has been involved over the past several years in an
on-going program in which the exhaust effluents from Titan rockets are
measured and characterized routinely to assess possible environmental
effects. Measurements are made *in situ* in the "stabilized ground
cloud" formed just after the launching of the rocket from aboard a
small aircraft (Wornon *et al.*, 1977). The SEM photomicrographs in
Figures 3A-3D show particles collected in the QCM cascade impactor
from the Titan III exhaust cloud. The particles in (A) were collected
on stage 7 of the impactor. The x-ray energy spectra for the spherical shaped ones show only aluminum and chlorine. They are presumed
to be Al_2O_3 particles formed from the burning of the fuel with a
coating of HCl. The HCl is also a combustion product. Since most of
these particles are spherical in shape they may be ideal for scattering measurements. However, there are a few non-spherical particles
on stage 7, and the particles collected in other stages of the cascade
are not regular in shape either; this fact presents problems in interpretation of their scattering characteristics (Chaun and Woods,
1978). (B) shows a 3.2×10^4 square micrometer area containing particles collected on stage 8. There are several large crystal-like
particles present along with a huge number of agglomerates. (C) shows
a higher magnification of a section of this same area. It now becomes clear that there are three types of morphologies. There are
small spherical singlets of about the right size for stage 7, there
are the agglomerates, and the large crystalline-like particles which
have both a diamond shape and the long rectangle shape as in (D).
The x-ray energy spectra show that the diamond-shaped particles consist of calcium and sulfur which could be a form of calcium sulfate.
The rectangular-shaped crystals show no peaks in the x-ray spectra
which means that they consist of elements with atomic numbers less
than 11.

5. VOLCANIC PLUME

In February, 1978 the QCM cascade impactor was flown aboard the
NCAR Queen Aire during a mission in which the effluents from several
active volcanos near Guatemala were sampled. The aircraft flew
through the plumes at different downwind distances from the volcano.

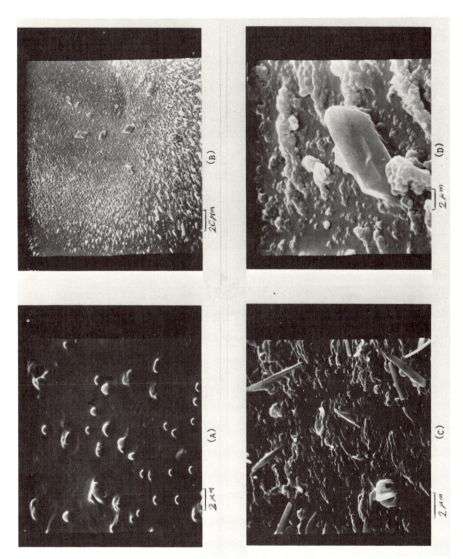

FIGS. 3A-3D. Particles collected in QCM stages 7, 8, and 9 from the exhaust plume of a Titan III rocket launched in August, 1977.

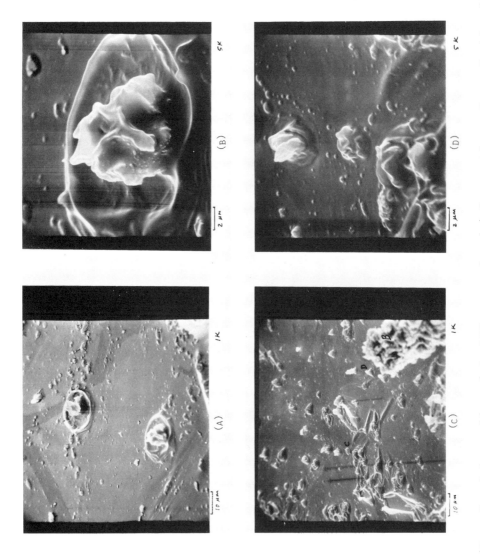

FIGS. 4A–4D. Particles collected in QCM stages 3 and 4 from active volcano Santiaguito in February, 1978.

Some particles collected are shown in Figures 4A-4D. The particles collected on stage 3 are shown in Figure 4(A) and 4(B). The particles consist of small nearly spherical droplets and a few large irregular shaped particles on the order of about 5 to 8 μm. These larger particles appear to have been coated with a liquid which splattered and formed a pool surrounding the particle upon impact on the surface of the crystal. (B) shows a higher magnification of one of these particles. Such a particle immersed in a liquid would undoubtedly produce a scattering pattern different from that of a spherical particle of the same size due to a mixed index of refraction. (C) and (D) show particles collected on stage 4. The particles in (C) appear to be relatively dry compared to the ones in stage 3. There are a few large ones that are more crystalline-like in shape plus a larger number of amorphous ones. Clearly these particles with their jagged edges would produce complicated scattering.

6. CONCLUSION

The examples presented here represent particles from several locations and sources. They illustrate the existence of a substantial number of non-spherical particles in various kinds of plumes and in the ambient atmosphere, both stratosphere and troposphere. These examples lead one to question the assumption of spherical particles for use in various calculations and instruments. Therefore, more attention must be given to the fact that a large percentage of atmospheric aerosols are non-spherical when using scattering data to infer size information or when performing calculations of their effects. Furthermore, when these particles differ from one another in composition and when some of them are immersed in liquid, there are additional complexities.

REFERENCES

Chuan, R. L., 1976, *in* "Fine Particles, Aerosol Generation, Measurement Sampling, and Analysis," Benjamin Y.H. Liu, Ed., Academic Press, New York.

Chuan, R. L. and Woods, D. C., 1978, *in* "Proceedings of the 4th Joint Conference on Sensing of Environmental Pollutants," American Chemical Society.

Junge, C. E., 1963, "Air Chemistry and Radioactivity," Academic Press, New York, p. 196.

Wornon, D. E., Woods, D. C., Thomas, M. E., and Tyson, R. W., 1977, *in* "Instrumentation of Sampling Aircraft for Measurement of Launch Vehicle Effluents," NASA TMX-3500.

ABSORPTION BY SMALL REGULAR NON-SPHERICAL PARTICLES

IN THE RAYLEIGH REGION

Janon Embury

U.S. Army Chemical Systems Laboratory

Aberdeen Proving Ground, Maryland   21010

ABSTRACT

Smaller particles characterized by a complex refractive index generally absorb more radiation per unit mass than larger particles. This absorption reaches a maximum in the Rayleigh region where it becomes size independent. We explore the dependence of dipolar resonant absorption on both shape and refractive index in the Rayleigh region.

Extinction by particles small compared to the incident radiation wavelength both inside and outside the particle is described by the Rayleigh approximation. Here extinction is dominated by absorption when the refractive index has an imaginary component. Absorption is independent of size as long as the approximation remains valid. The absorption properties of a particle are not only a function of complex refractive index but are also very strongly dependent on shape. Unlike many other size and refractive index regions where shape effects are difficult to predict, a good description of shape effects is provided by the Rayleigh ellipsoidal approximation which predicts the absorption and scattering properties of small ellipsoidal particles (van de Hulst, 1957).

The absorption coefficient for a collection of randomly oriented Rayleigh ellipsoids is:

$$\text{Absorption Coefficient} = \frac{\langle\text{Optical Absorption Cross Section}\rangle}{\text{Weight}} = \frac{\langle C_{abs}\rangle}{\text{Weight}}$$

97

where the brackets represent an average over all particle orientations. The optical absorption cross section is defined as

$$<C_{abs}> = <\text{Geometric Cross Section x Absorption Efficiency}> .$$

The average optical absorption cross section at wavelength λ depends
on the polarizability α_j of the ellipsoidal particle along each of
the three semiaxes (Kerker, 1969):

$$<C_{abs}> = \frac{2\pi}{\lambda} \text{Im}\{\frac{1}{3} \sum_{j=1}^{3} \alpha_j\}$$

where

$$\alpha_j = \frac{V(\varepsilon/\varepsilon_m - 1)}{L_j(\varepsilon/\varepsilon_m + 1)} ,$$

$\varepsilon = \varepsilon_1 + i\varepsilon_2$ = dielectric constant of particle,

ε_m = dielectric constant of medium supporting particles
$(\varepsilon_m = 1$ for air),

V = particle volume,

L_j = depolarization factor along semiaxis j .

The three depolarization factors have the properties that each is
nonnegative and they sum to unity. The optical absorption cross
section now takes the form:

$$<C_{abs}> = \frac{2\pi V}{\lambda} \sum_{j=1}^{3} \frac{\varepsilon_m \varepsilon_2}{[L_j\varepsilon_1 + \varepsilon_m(1-L_j)]^2 + (L_j\varepsilon_2)^2} .$$

When the shape and dielectric properties of the particle combine to
reduce the value of the denominator, there is an enhancement in
absorption or equivalently a resonance in absorption (Huffman, 1977).
The shape dependence of this resonance enters through the depolarization factor. A sphere has depolarization factors, L_j, equal to
1/3 for fields applied along any orthogonal set of three radial axes.
A prolate spheroid approximating a thin needle has $L_j \to 0$ for fields
applied parallel to its length and $L_j \to 1/2$ for fields in the plane of
symmetry. An oblate spheroid approximating a thin disc has $L_j \to 1$ for
fields applied perpendicular to the plane of symmetry and $L_j \to 0$ for
fields in the symmetry plane.

A resonance occurs when the following conditions are satisfied:

$$L_j \varepsilon_1 + \varepsilon_m (1-L_j) = 0 \; , \; L_j \varepsilon_2 = 0 \; .$$

For a given shape and depolarization factor, the resonant value for ε_1 becomes

$$\varepsilon_1 = \varepsilon_m (1 - 1/L_j) \; .$$

This resonance condition specifies only negative values for ε_1 as a result of the fact that the depolarization factor must assume values between zero and one. The second condition for a resonance is satisfied for nonzero depolarization factors when $\varepsilon_2 = 0$.

Translating these two resonant values for dielectric constant into resonant values for the optical constants we use the relationships

$$\varepsilon_1 = n^2 - k^2 , \quad \varepsilon_2 = 2nk$$

where n = real part of the complex refractive index, and k = imaginary part of the complex refractive index. The optical constant resonant values then become

$$n = 0, \quad k = \{ \varepsilon_m (1/L_j - 1) \}^{1/2}$$

All three depolarization factors for a sphere are equal to 1/3 and its resonance in air ($\varepsilon_m = 1$) occurs at n = 0 and k = $\sqrt{2}$. In Figure 1, contour values for extinction in $m^2 \, g^{-1}$ are shown for a sphere at a wavelength of one micron and density of one g cm^{-3}. As the shape becomes progressively more prolate, one pole moves up the k axis and the other converges on a value of 1 on the k axis. In the case of an oblate spheroid, one pole moves up the k axis and the other converges on the origin as it becomes flatter. This is evident in Figure 2 where B is the spheroid's axis of symmetry. If the region in the complex refractive index plane around the origin were magnified we would be able to see large absorption contours very close to the origin.

The special case where $L_j \rightarrow 0$ warrants some discussion. The depolarization factor approaching zero occurs for selected orientations of thin needles and thin discs with respect to the electric field vector for incident radiation. The electric field vector must lie parallel to the length of a thin needle shaped particle or lie within the plane of symmetry of a thin disc shaped particle. Both

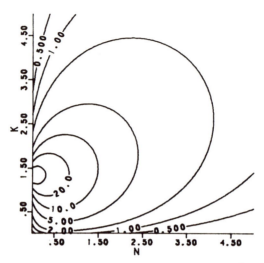

FIG. 1.   Isopleths of constant extinction (m^2 g^{-1}) for a unit density
Rayleigh sphere at a wavelength of 1 micron.

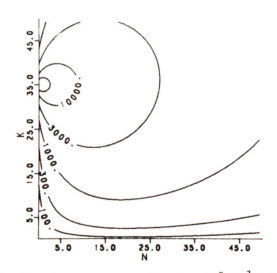

FIG. 2.   Isopleths of constant extinction (m^2 g^{-1}) for a unit density
Rayleigh *oblate* spheroid with aspect ratio 0.001 at a wavelength of
1 micron.

of these shapes can be approximated by appropriately elongated prolate and flattened oblate spheroids respectively.

The sensitivity of resonance strength to changes in ε_1, and ε_2
away from resonance is minimized as the depolarization factor
approaches zero. This is evident when we write down the absorption
cross section when the resonant dominant term(s) involve a depolarization factor approaching zero and ε_1 deviates by δ_1 away from its
resonance value and ε_2 remains nonzero:

$$<C_{abs}> = \frac{2\pi V}{\lambda} \sum_{\substack{resonant \\ L_j \to 0}} \frac{\varepsilon_m \varepsilon_2}{L_j^2(\delta_1^2+\varepsilon_2^2)} \quad .$$

In the limit where the depolarization factor is nearly zero, it is
clear that both δ_1 and ε_2 can become large and the absorption cross
section will remain large. To maintain strong absorption as L_j becomes small, all that is required is that

$$\frac{\varepsilon_2}{\delta_1^2+\varepsilon_2^2} >> L_j^2 \quad \text{with} \quad L_j \to 0 \ .$$

Now there is much greater latitude in the choice of ε_1 and ε_2 that
will produce strong absorption. Consequently it is much easier to
find materials producing high extinction when their shape results in
a depolarization factor of zero for at least one orientation.

The limiting case where $L_j=0$, although never realizable in
nature, causes the absorption cross section to take the following
form

$$<C_{abs}> = \frac{2\pi V}{\lambda} \sum_{L_j=0} \varepsilon_2/\varepsilon_m + \text{terms with } L_j \neq 0 \ .$$

It is clear now that the earlier resonance conditions on ε_1 are irrel-
evant since ε_1 does not appear in this expression. The highest optical absorption cross sections are encountered when ε_2 becomes large
or equivalently when the product nk becomes large. This is evident
in Figure 3 where contours of constant absorption cross section in
m^2 g^{-1} appear at a wavelength of one micron and a density of $1g$ cm^{-3}.
A prolate spheroid with an aspect ratio very far from one, in this
case the symmetry axis is A, has a depolarization factor along its
length very close to zero. As evidenced by the contour lines there
is an insignificant contribution from the resonant terms corresponding to the other two depolarization factors which don't approach

zero throughout most of the complex refractive index plane. If we
were to magnify the part of the refractive index plane close to
n = 0 and k = 1 we would find large absorption contours, but in a
much more confined region.

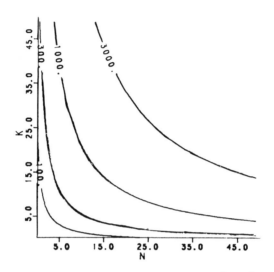

FIG. 3.   Isopleths of constant extinction (m^2 g^{-1}) for a unit density
Rayleigh *prolate* spheroid with aspect ratio 0.001 at a wavelength of
1 micron.

Question (GREENBERG): Your use of Rayleigh scattering depends on the
validity of the Rayleigh approximation. For high refractive indexes,
that produces a severe limitation on how long your fibers can be.

Answer (EMBURY):   True; however, the parametric contour plots
shown were based only on the length/diameter ratio. With high refractive indexes you may find that some regions of the n, k plane
will require finding extremely thin particles to show Rayleigh
extinction.

REFERENCES

Huffman, D. R., 1977, *Advances in Physics*, 26, 129.
Kerker, M., 1969, "The Scattering of Light and Other Electromagnetic
   Radiation," Academic, N. Y.
van de Hulst, H. ., 1957, "Light Scattering by Small Particles,"
   John Wiley and Sons, New York.

INFRARED ABSORPTION SPECTRA OF NON-SPHERICAL PARTICLES

TREATED IN THE RAYLEIGH-ELLIPSOID APPROXIMATION

Donald R. Huffman

Univ. of Arizona, Dept. of Physics

Tucson, Arizona 85721

Craig F. Bohren

Univ. of Arizona, Institute of Atmospheric Physics

Tucson, Arizona 85721

ABSTRACT

A comparison of small particle extinction spectra calculated for spheres with extinction measurements shows the inadequacy of the spherical approximation in treating strong infrared absorption bands. A simple expression for absorption cross sections of non-spherical particles is derived by integrating over a distribution of shape parameters in the Rayleigh-ellipsoid approximation. For comparison with the theoretical results infrared extinction measurements were made of sub-micron quartz particles dispersed in a KBr matrix. The favorable comparison of theory and experiment suggests that the treatment of shape effects in infrared absorption spectra by means of a distribution of ellipsoids constitutes a major improvement.

1. INTRODUCTION

There is a common feeling that particles small enough to be described by Rayleigh theory do not show very strong shape effects, especially in absorption. This is not at all true, however, in the case of strong absorption bands in the infrared spectra of small particles. In spectral regions where the real part of the complex dielectric function is strongly negative, small particle extinction

103

bands are markedly shifted and broadened as compared to the corresponding bulk solid bands. The shift is understandable from the
theory for spheres, but the broadening has been more difficult to
explain, although it has been felt that a distribution of shapes
would provide the necessary broadening mechanism (Gilra, 1972;
Huffman, 1977). Figure 1 illustrates the problem with the example
of crystalline quartz. The sphere calculations are based on measured
optical constants (Spitzer and Kleinman, 1961). Extinction measurements are for sub-micron crystalline quartz particles dispersed in a
KBr matrix. Details of the measurements will be given later in the
paper. The point we are making now is that sphere calculations do
not satisfactorily predict the magnitude and shape of infrared lattice
bands in real, non-spherical particles. The calculated maximum

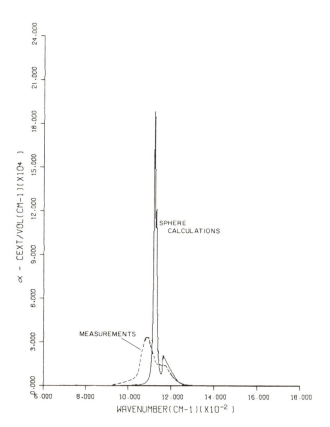

FIG. 1.   Comparison of volume normalized extinction calculations for
spheres with extinction measurements of small particle quartz dispersed in a KBr matrix. Calculations are based on measured dielectric functions of Spitzer and Kleinman (1961), relative to KBr.

extinction is incorrect by almost an order of magnitude, and the
shape of the band is poorly predicted by the theory.

   In this paper we treat such effects by integrating over all
ellipsoidal shape parameters in the Rayleigh-ellipsoid approximation
to obtain a simple expression for the absorption cross section per
unit volume of randomly oriented, non-spherical particles. This idea
derived from conversation with D.P. Gilra who did calculations for
a distribution of *spheroids*. Gilra's unpublished calculations were
used by Treffers and Cohen (1974) in an attempt to identify particles
in the space surrounding cool stars by means of features in the infrared emission spectra. Our calculations, which use measured optical
constants, are compared with laboratory extinction measurements on a
common particulate system in order to show that the simple theory is
much more satisfactory for handling real particle shape effects than
the sphere. calculation.

2. THEORY OF SHAPE EFFECTS IN THE RAYLEIGH-ELLIPSOID APPROXIMATION

   The average absorption cross section for a collection of randomly oriented identical homogeneous ellipsoids that are sufficiently
small compared to the wavelength of the incident light may be written
in the form (van de Hulst, 1957, chap. 6)

$$<C_{abs}> = kV/3 \ \mathrm{Im} \ \{ \sum_{j=1}^{3} (\beta+L_j)^{-1} \} \ , \tag{1}$$

$$\beta = (\epsilon-1)^{-1} \ ,$$

where $V = 4\pi abc/3$ is the volume of an ellipsoid with semiaxes of
length $a \geq b \geq c$; $\epsilon = \epsilon' + i\epsilon''$ is the complex dielectric function
of the particle relative to that of the surrounding medium, and
$k = 2\pi/\lambda$ is the wave number of the surrounding medium. The L_j's are
geometrical factors related to the ratios of semi-major axes for the
ellipsoid (van de Hulst, 1957, p. 71). Only two of the geometrical
factors, which we shall take to be L_1 and L_2 are independent because
of the relation $L_1 + L_2 + L_3 = 1$.

   Suppose now that, in addition to being randomly oriented, the
collection consists of all possible ellipsoidal shapes; *i.e.*, the
geometrical factors are not restricted to a single set of values but
are distributed according to some shape probability function $P(L_1,L_2)$.
All ellipsoidal shapes are represented by points in the shaded triangular region shown in Figure 2. However, it is convenient for
purposes of integration to define P on the larger triangular region
Γ, which is composed of six equal area regions, each of which corresponds to one of the six possible ways of choosing the relative
lengths of the ellipsoid axes a, b, c. The shape probability function

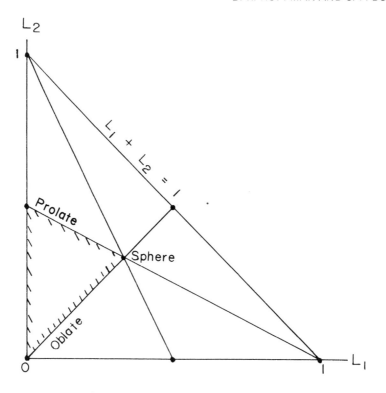

FIG. 2.   Domain of definition of the shape probability function
P(L_1,L_2).   Outside of the region Γ bounded by L_1 = 0, L_2 = 0,
L_1 + L_2 = 1,   P(L_1,L_2) vanishes identically.

is normalized to unity on Γ:

$$\iint_{\Gamma} P(L_1,L_2) \; dL_1 \; dL_2 = 1 \; . \tag{2}$$

The absorption cross section averaged over the shape distribution and
over all orientations is, therefore,

$$
\begin{aligned}
\langle\langle C_{abs}\rangle\rangle &= \iint \langle C_{abs}\rangle \; P(L_1,L_2) dL_1 dL_2 \\
&= kV/3 \; \text{Im} \; \{ I_1 + I_2 + I_3 \} \; ,
\end{aligned}
\tag{3}
$$

where

$$I_1 = \iint \frac{P(L_1,L_2)}{\beta + L_1} \; dL_1 \; dL_2 \; ,$$

$$I_2 = \iint \frac{P(L_1,L_2)}{\beta + L_2} \; dL_1 \; dL_2 \; ,$$

$$I_3 = \iint \frac{P(L_1,L_2)}{\beta + 1 - L_1 - L_2} \; dL_1 \; dL_2 \; ,$$

and $\beta = (\varepsilon-1)^{-1}$. The integral of a function $f(L_1,L_2)$ over Γ may be written as an iterated integral:

$$\iint_\Gamma f(L_1,L_2) dL_1 \; dL_2 = \int_0^1 dL_1 \int_0^{1-L_1} f(L_1,L_2) \; dL_2 \; .$$

We have assumed that all particles have the same volume V; however, if there is no correlation between shape and volume, then the total absorption cross section of the collection is

$$\eta \frac{k<V>}{3} \; Im \; \{I_1 + I_2 + I_3\}$$

where η is the total number of particles per unit volume and $<V>$ is the average particle volume. It has also been implicitly assumed that $P(L_1,L_2)$ is continuous; this is not a necessary restriction, however, and we can take into account discrete distributions by replacing the above integrals with summations over the discrete set of points (L_1,L_2) in Γ.

Perhaps the simplest conceivable distribution is one for which all shapes are equally probable, in the sense that $P(L_1,L_2) = 2$, and the integrals are readily evaluated

$$I_1 = I_2 = I_3 = \frac{2\varepsilon}{\varepsilon-1} \; Log\varepsilon - 2 \; .$$

Therefore, the average cross section is

$$<<C_{abs}>> = k<V> \; Im \left(\frac{2\varepsilon}{\varepsilon-1} \; Log\varepsilon \right) \; . \tag{4}$$

In the previous expression Log denotes the *principal value* of the logarithm of a complex number $z = re^{i\theta}$ (Churchill, 1960, p. 56):

$$\text{Log } z = \log r + i\theta \quad (r > 0, -\pi < \theta < \pi).$$

Other functions $P(L_1, L_2)$ could be chosen to represent a shape distribution of particles. The one chosen (equation (2)) has the advantage of uniformity in the $L_1 L_2$ plane of Figure 2 and simplicity of the resulting expression, equation (4). The real test of its practical value, of course, lies in how well it describes the properties of actual particulates.

3. COMPARISON OF THEORY AND EXPERIMENT

Crystalline quartz has been chosen for illustration of the results of equation (4) as compared with actual measurements. Quartz is anisotropic, but both sets of optical constants have been measured by Spitzer and Kleinman (1961). In using equation (4) for this anisotropic solid it has been necessary for us to assume that the randomization of crystal axes with respect to the ellipsoid axes will be properly accomplished by weighting optical constants for the ordinary ray by 2/3 and for the extraordinary ray by 1/3. Only in the particular case of spheres has this been shown to be correct, and such weighting was used in the sphere calculation of Figure 3. Although we have not been able to prove this weighting procedure rigorously, we have a strong feeling that this is at least close to being correct. Randomization of the ellipsoid directions *in space* and randomization of the ellipsoid *shape parameters* have both been accounted for rigorously. Favorable comparison of theory and experiment is in some sense the justification for the procedures.

Samples for transmission measurements were prepared by dispersing a powder of crystalline SiO_2 in water and allowing it to settle for about 18 hours. The top portion of the liquid, containing sub-micron particles, was then drawn off and gently heated overnight to dry the particles. Small portions of the segregated powder were weighted and mixed with KBr powder by agitation in a glass vial with steel balls overnight. The mixture was pressed with a ten ton press into a 1/2" diameter pellet for infrared transmission studies.

The infrared transmission spectrum was recorded with a Perkin Elmer 137 spectrophotometer with a compensating blank of pure KBr in the reference beam. Extinction coefficients (α) normalized per unit volume of sample solid were calculated from transmission (T) measurements using the equation

$$T = \exp\left(-\alpha \frac{\sigma}{\rho}\right)$$

where σ is the mass of particles per unit cross sectional area, and ρ is the bulk density. Since the particles are quite small compared to the wavelength, scattering is negligible and extinction is very nearly equal to absorption.

Extinction cross section results are plotted in Figure 3 along with theoretical calculations using measured optical constants. Both

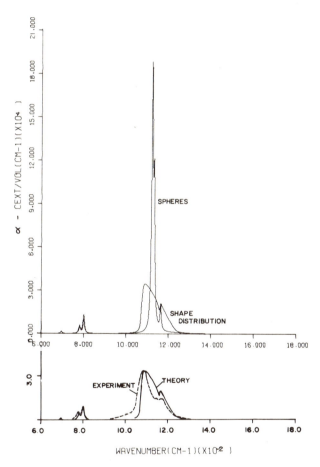

FIG. 3.   Extinction cross sections normalized per unit volume of bulk solid for small particles of quartz. The upper curves compare cal-culations for spheres and for a distribution of ellipsoidal shapes (equation (4)). The lower curve shows results from measurements taken on quartz particles embeded in a KBr matrix. Measured extinc-tion is determined from transmission measurements and mass measure-ments. There is no arbitrary normalization in either experiment or theory.

sets of calculations, for spheres and for the ellipsoidal shape distribution of equation (4), are presented in the upper curves. The
lower curves reproduce the ellipsoid distribution calculations (solid
lines) along with the measurements (dashed lines) for ready comparison. Measurements shown are for one of eight quartz samples measured.
Although the extinction curves have not been averaged, this particular sample was chosen because its peak extinction value is within 2%
of the average peak extinction for the nine different samples whose
average deviation from the mean was 8%. There is no arbitrary normalization of either the theoretical or the experimental curves. The
superiority of the shape distributions over the sphere calculations
is clearly seen. Maximum extinction is very well predicted by the
theory, and the shape of the theoretical bands compares favorably
with the measurements.

Of course, real particles are no more ellipsoidal than they are
spherical. Apparently the reason our idealized representation of
shape distribution is so satisfactory is that it provides a simple
averaging over all possible resonant modes that occur in the negative
ϵ' region. Irregular particles have a different distribution of such
modes. Fuchs (1975), for example, has calculated the positions and
strengths of such modes in a cubic particle, and Langbein (1976) has
determined their positions for a variety of rectangular parallelepi-
peds. These rectangular particles give narrow resonance bands
throughout the region of negative ϵ'. When the particles are irregular enough that the entire negative ϵ' region is populated with the
modes, the result for real particles apparently is quite similar to
the population of all resonant modes occurring in our random collection of ellipsoids.

4. CONCLUSIONS

Shape effects are very important in determining the strengths
and shapes of strong infrared absorption bands in small particles.
As a result the extinction spectra calculated for spheres poorly
approximates real extinction spectra such as are determined from a
dispersed powder in a KBr matrix. Our new theoretical expression,
which accounts for a continuous distribution of shapes in the
Rayleigh-ellipsoid approximation, gives a much more satisfactory representation of small quartz particles than does the sphere theory.
Both the strengths and the shapes of absorption bands are accurately
predicted.

Acknowledgements. This paper is based on a section in a forth-
coming book, "Absorption and Scattering of Light by Small Particles"
by C. F. Bohren and D. R. Huffman, to be published by Wiley-Inter-
science in 1980. We are indebted to Jan Rathmann and Lin Oliver for
their help with the experimental measurements and with the data analysis. The work was supported by the U.S. Naval Air Systems Command.

Comment (ARONSON): We used the same kind of treatment to estimate the particle size of the Martian dust observed by the NASA IRIS instrument aboard Mariner 9. NASA had estimated a size between 2 and 20 μm by Mie theory calculations, but using a distribution of ellipsoids such as you gave, we inferred a slightly submicron value by comparing infrared bandwidths observed by the IRIS instrument with theoretical calculations. When we told NASA of our result, they said that it was in agreement with observations of the settling rate.

REFERENCES

Churchill, R. V., 1960, "Complex Variables and Applications," McGraw-Hill, N.Y.

Fuchs, R., 1975, *Phys. Rev. B11*, 1732.

Gilra, D. P., 1972, "The Scientific Results From the Orbiting Astronomical Observatory OAO-2," A. D. Code, Ed., (NASA SP-310) p. 295; Ph.D. thesis, University of Wisconsin.

Huffman, D. R., 1977, *Advances in Physics, 26*, 129.

Langbein, D., 1976, *J. Phys. A: Math. Gen., 9*, 627.

Spitzer, W. G., and Kleinman, D. A., 1961, *Phys. Rev., 121*, 1324.

Treffers, R. and Cohen, M., 1974, *Astrophys. J., 188*, 545.

van de Hulst, H. C., 1957, "Light Scattering by Small Particles," Wiley, N.Y.

SCATTERING BY NON-SPHERICAL PARTICLES OF SIZE COMPARABLE

TO A WAVELENGTH: A NEW SEMI-EMPIRICAL THEORY

James B. Pollack and Jeffrey N. Cuzzi

NASA-Ames Research Center, Space Science Division

Moffett Field, California 94035

ABSTRACT

We propose an approximate method for evaluating the interaction of randomly oriented, non-spherical particles with the total intensity component of electromagnetic radiation. When the particle size parameter, x, the ratio of particle circumference to wavelength, is less than some upper bound x_0 (~ 5), Mie theory is used. For $x > x_0$, the interaction is divided into three components: diffraction, external reflection, and transmission. Physical optics theory is used to obtain the first of these components; geometrical optics theory is applied to the second; and a simple parameterization is employed for the third. The predictions of this theory are found to be in very good agreement with laboratory measurements for a wide variety of particle shapes, sizes, and refractive indexes. Limitations of the theory are also noted.

1. INTRODUCTION

We have attempted to formulate a systematic method for treating the *scalar* scattering properties of *randomly oriented* non-spherical particles. Thus, our attention is directed toward the total intensity of the scattered light. We have developed a semi-empirical theory which is based on simple physical principles and comparisons with laboratory measurements. The ultimate utility of our procedure rests on its ability to successfully reproduce the observed single scattering phase function for a wide variety of particle shapes, sizes, and refractive indexes. Our methods and results are discussed in more detail by Pollack and Cuzzi (1979).

2. EXISTING EXPERIMENTAL DATA

Two types of laboratory data exist: *direct measurements* of scattering by ensembles of particles at visible wavelengths and *analog experiments* that perform similar investigations on individual particles at microwave wavelengths, with the results combined mathematically to simulate a randomly oriented, polydispersed system.

We first consider laboratory measurements of the extinction efficiency, Q_{ext} (*e.g.* van de Hulst, 1957). Figure 1 illustrates the dependence of Q_{ext} on the phase shift parameter $\rho = 2 x (m_r-1)$ for rough, approximately equidimensional dielectric particles that are composed of stacked cylinders (Greenberg *et al.*, 1971), where x, a, m_r, and λ are, respectively, the size parameter $2\pi a/\lambda$, radius of an equal volume sphere, real part of the index of refraction m, and wavelength. The vertical lines in Figure 1 represent the range of measured values of Q_{ext} for a fixed value of ρ. The triangles show the relationship between Q_{ext} and ρ for equal volume spheres, as given by Mie theory.

Several conclusions can be drawn from Figure 1. First, for $\rho \lesssim 4$, the stacked cylinders have about the same dependence of Q_{ext} on ρ as do equal volume spheres. Second, for $\rho > 4$, Q_{ext} is systematically higher than the value appropriate for the equivalent sphere. This disagreement is not unexpected, as discussed in Sections 3.2 and 3.5. Other laboratory measurements on irregular particles (Berry, 1962; Proctor and Barker, 1974; Proctor and Harris, 1974; Greenberg *et al.*, 1961) support the above conclusions and extend them to larger values of ρ.

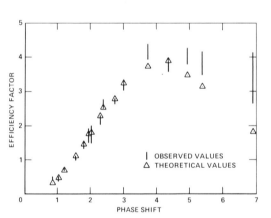

FIG. 1. Extinction efficiency as a function of the phase shift parameter ρ for particles composed of stacked cylinders. The vertical lines show these results for different orientations of the electric vector (Greenberg *et al.*, 1971). The triangles are the Mie scattering predictions for equal volume spheres having the same refractive indexes. The phase shift parameter is defined in the text.

For irregular particles with ρ less than some critical value ρ_0, where ρ_0 is a function of shape but is typically ~3 to 10, the phase function also agrees well with that of equal volume spheres as shown in Figures 2a and 2b. However, when $\rho > \rho_0$, non-spherical particles show marked deviations in their phase functions from that of

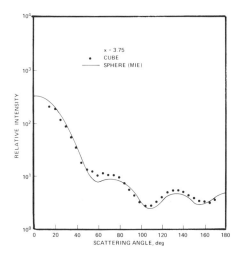

FIG. 2a.   Comparison of the angular distribution of scattered intensity obtained using analog techniques (dots) with Mie theory for an equal volume sphere having the same refractive index (solid line) (after Zerull and Giese, 1974) for a small cube.

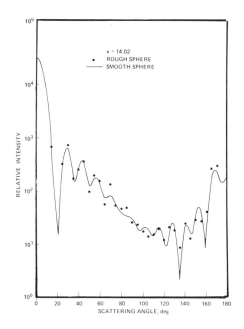

FIG. 2b.   Same as Fig. 2a but for a corrugated sphere.

their spherical counterparts, especially at intermediate and large angles of scatter. This point is illustrated in Figures 3 and 4. In both figures, resonant features, such as the Mie backscatter peak, are absent in the data for the non-spherical particles. We also note that at those angles where there is a sizeable departure from Mie scattering behavior, the logarithm of the phase function varies approximately linearly with scattering angle.

Figures 3 and 4 also show that Mie scattering theory provides an approximate fit to the measurements at small angles of scatter. Further evidence for this is most clearly given by the visible light scattering experiments and calculations of Hodkinson (1963). Hodkinson also approximated the phase function of particles with $x \sim 10-100$ by the sum of the diffraction pattern of an opaque disk of equal projected area and the external reflection and internal refraction components calculated according to geometrical optics theory. Hodkinson and Greenleaves (1963) had previously shown that this recipe gave a good fit to Mie theory results at small scattering angles for particles with $x \sim 10-25$.

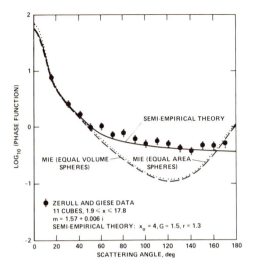

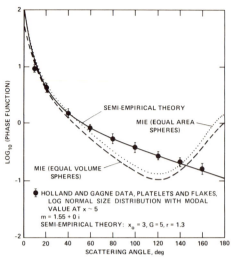

FIG. 3.   Comparison of the phase
function for scattering by a particular size distribution of cubes
(dots with error bars), as obtained by numerically integrating
results from analog measurements
(Zerull and Giese, 1974), with
the predictions of Mie theory for
equal volume and equal area spheres
and the semi-empirical theory of
this paper.   Each theoretical
curve has been normalized to its
correct vertical position regarding total scattered energy
with respect to the data points.
Consequently, the normalization
convention differs slightly among
the three theoretical curves.

FIG. 4.   Comparison of the
measured phase function for
scattering by an ensemble of
flat plates (dots with error
bars; Holland and Gagne, 1970),
with the predictions of Mie
theory for equal volume and
equal area spheres and with the
semi-empirical theory of this
paper.

3.   A SEMI-EMPIRICAL THEORY

Our treatment of light scattering by randomly oriented, irregularly shaped particles is a combination of Mie theory, physical
optics, geometrical optics, and parameterization.   When x is less
than some upper bound x_0, we use the Mie cross sections and phase
function directly.   For larger particles, appropriately scaled Mie
theory results are used to define the cross sections, and the phase
function is constructed from the sum of three components -- those due
to diffraction, external reflection, and internal transmission.

3.1   Small-Particle Regime

The small particle regime is defined by the criterion that $x = 2\pi a/\lambda \leq x_0$, where x_0 is a parameter to be obtained from experimental results. Typically, $x_0 \sim 5$. Within this size regime, Mie scattering theory is used to determine the phase function and the efficiencies for absorption, scattering, and extinction. The laboratory data discussed in the previous section provide the basic justification for the use of Mie theory in the small size regime (cf. Figs. 1 and 2). Note that these data indicate that this approximation is appropriate for sizes up to $x = x_0 \gtrsim 1$.

Possible limitations on the applicability of our prescription can be ascertained from theoretical results (Jones, 1972; Greenberg, 1972; Atlas *et al.*, 1953) obtained for randomly oriented spheroids when $x \ll 1$. For particles with $(|m^2|-1) \lesssim 1$, the fractional deviation from Mie theory is only $\sim 10\%$ even for infinitely elongated particles. For $(|m^2|-1) \gtrsim 10$, the deviation is a factor of several. Thus, use of Mie theory for the small size regime loses its validity for very elongated particles, particularly ones having large values of $(|m^2|-1)$.

3.2   Large-Size Regimes: Diffraction Component

The diffracted component is assumed to be that of an opaque circular disk having an area equal to the irregular particle's projected area. The latter is rigorously equal to one-fourth the irregular particle's total surface area for convex (non-reentrant) particles in random orientation (Vouk, 1949). Since irregular particles always have greater surface areas than that of equal volume spheres by some factor r, the radius of the disk $\tilde{a} = a\sqrt{r}$. Following Hodkinson and Greenleaves (1963), we have approximated the phase function for the diffraction component, I_D, by

$$I_D = C_D \frac{(\tilde{x})^2}{4\pi} \left[\frac{2J_1(z)}{z} \right]^2 k \simeq \frac{2C_D \sin^2(\tilde{x}\sin\Theta - \pi/4)k}{\pi^2 \tilde{x}\sin^3\Theta} \qquad (1a)$$

where

$$z = \tilde{x}\sin\Theta = \frac{2\pi\tilde{a}}{\lambda}\sin\Theta \quad , \quad k = \frac{1}{2}(1 + \cos^2\Theta) \quad .$$

J_1 is the first order Bessel function, Θ is the angle of scatter, and C_D is obtained from the usual normalization condition by integrating over solid angle,

$$\int I_D \frac{d\Omega}{4\pi} = 1 \quad . \tag{1b}$$

The diffraction contribution to the extinction efficiency, Q_D, equals 1, in accord with Babinet's principle.

3.3 Large-Size Regime: External Reflection Component

An ensemble of randomly oriented convex, irregularly shaped particles reflects light from their external surfaces into the same angular distribution as does an ensemble of equal area spheres with the same refractive indexes (van de Hulst, 1957; Hodkinson, 1963; Hansen and Travis, 1974). We assume that the geometrical optics expressions for external reflection can be applied to all particle sizes in the large size regime, $i.e.$, $x > x_0$. The phase function for external reflection by irregular particles, I_R, is obtained by averaging the Fresnel reflection coefficients for orthogonal polarizations (Hodkinson and Greenleaves, 1963):

$$
\begin{aligned}
I_R = \frac{1}{2} C_R & \left[\frac{\sin(\Theta/2) - \left(|m|^2 - 1 + \sin^2(\Theta/2) \right)^{1/2}}{\sin(\Theta/2) + \left(|m|^2 - 1 + \sin^2(\Theta/2) \right)^{1/2}} \right]^2 \\
+ \frac{1}{2} C_R & \left[\frac{|m|^2 \sin(\Theta/2) - \left(|m|^2 - 1 + \sin^2(\Theta/2) \right)^{1/2}}{|m|^2 \sin(\Theta/2) + \left(|m|^2 - 1 + \sin^2(\Theta/2) \right)^{1/2}} \right]^2
\end{aligned}
\tag{2a}
$$

where

$$|m|^2 = m_r^2 + m_i^2 \quad ;$$

m_r and m_i are the real and imaginary parts of the index of refraction, and C_R is found from:

$$\int I_R \frac{d\Omega}{4\pi} = 1 \quad . \tag{2b}$$

The efficiency factor for external reflection, Q_R, is related to C_R by

$$Q_R = 1/C_R \quad . \tag{3}$$

Note that both I_R and Q_R are independent of particle size.

Justification for our use of equations (2) and (3) for sizes only somewhat larger than x_0, *i.e.*, below the usual boundary for geometrical optics ($x \gg 1$), is given by the studies of Hodkinson (1963). His prescription, described in Section 2, gave a good fit to his measurements of scattering at angles less than about 60° by diamond dust with mean values of x ranging from 14 to 21. For angles greater than 30°, the diffraction component ceased being the dominant component, with the reflection component taking over this role because of the high relative refractive index of the diamond (1.8). At still larger angles of scatter, the internally transmitted component probably was the dominant one and responsible for the deviation between the observed and predicted phase functions.

3.4 Large-Size Regime: Transmitted Component

Some insight into the behavior of the transmitted component can be obtained by considering the geometrical optics limit, as illustrated in Figure 5 for cubes, flat plates, and spheres. As indicated in this figure, a substantial fraction of the radiation which enters one face of the cube is totally internally reflected at the adjacent, orthogonal face and emerges at a large angle of scatter. At the other extreme, flat plates with parallel large faces do not produce any deviation from the forward direction for most of the light entering the particle. In the case of spheres, some deviation in direction occurs, but, as pointed out by Hodkinson (1963), strong internal reflection occurs only for the small fraction of energy entering nearly tangentially, and total internal reflection is never attained. Thus, irregular particles may either be more or less effective than spheres at redistributing the incident, transmitted light into large scattering angles.

Rather than attempt the difficult task of determining the transmitted component in an exact manner, we have designed a parameterization which is as simple as possible and yet carries some physical significance. In accord with the observations discussed in Section 2, we assume the logarithm of the transmitted component's phase function, I_T, varies in a linear manner with scattering angle:

$$I_T = C_T \exp(-b\theta) \qquad (4a)$$

where C_T is found from

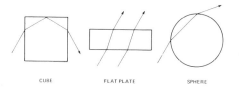

CUBE   FLAT PLATE   SPHERE

FIG. 5.   Paths of light rays through a cube, a flat plate, and a sphere.

$$\int I_T \frac{d\Omega}{4\pi} = 1 \quad , \qquad (4b)$$

and b is specified in terms of a second and more physically meaningful empirical constant G by

$$G \equiv \int_0^{\pi/2} I_T d\Theta \Big/ \int_{\pi/2}^{\pi} I_T d\Theta \ . \tag{4c}$$

G is similar to, but not quite the same as, the ratio of energy
scattered into the two hemispheres.

The scattering efficiency for the transmitted component, Q_T, is
given by $Q_T = Q_s^L - Q_D - Q_R$, where

$$Q_s^L = \int_{x_0}^{\infty} Q_s(x) n(x) \pi x^2 \ dx \Big/ \int_{x_0}^{\infty} n(x) \pi x^2 \ dx \ . \tag{5}$$

Q_s is the Mie scattering efficiency for a particle with size parameter
x and n(x) is the size distribution function for equal volume spheres.

3.5   Composite Phase Function and Efficiencies

We first consider the efficiency factors. We obtain the Mie
scattering and absorption efficiencies Q_s^L and Q_a^L for particles in the
large-size-regime using equation (5). Analogous equations, but with
0 and x_0 as the limits on the integral, are used to determine the Mie
scattering and absorption efficiencies Q_s^S and Q_a^S in the small-size
regime. At this point, we increase the scattering efficiency of the
large irregular particles relative to that of the small irregular
particles by a factor r as discussed in Sections 2 and 3.2. However,
the absorption efficiency in the large-size regime is proportional
to the volume of the scatterer, at least for $2m_1x < 1$, and is not so
affected. Thus, for the entire ensemble of irregular particles, the
composite scattering, absorption, and extinction efficiencies,
Q_s^*, Q_a^*, Q_e^*, are given by:

$$Q_s^* = Q_s^L \cdot F \cdot r + Q_s^S \cdot (1 - F) \tag{6a}$$

$$Q_a^* = Q_a^L \cdot F + Q_a^S \cdot (1 - F) \tag{6b}$$

$$Q_e^* = Q_s^* + Q_a^* \ , \tag{6c}$$

where

$$F = \int_{x_o}^{\infty} n(x)\pi x^2 \, dx \Big/ \int_{0}^{\infty} n(x)\pi x^2 \, dx \; . \tag{6d}$$

In cases where the particles are so highly absorbing that $2m_i x > 1$ in the large-size regime, their absorption cross section is proportional to surface area, and the first term on the right-hand side of (6b) should be multiplied by r.

The composite phase function, p^*, is obtained by summing the phase functions I_S, I_p, I_R, and I_T, with each one being weighted by its contribution to the total scattering:

$$p^* = I_S \cdot (1 - F) \cdot \frac{Q_S^S}{Q_s^*} + r \cdot F \cdot \frac{Q_S^L}{Q_s^*} \cdot$$

$$\left(I_D \cdot \frac{Q_D}{Q_S^L} + I_R \cdot \frac{Q_R}{Q_S^L} + I_T \cdot \frac{Q_T}{Q_S^L} \right) . \tag{7}$$

It can readily be shown that p^* fulfills the usual normalization condition.

3.6 Comparison with Laboratory Measurements

The ultimate test of the algorithm is how well it reproduces measurements of scattering behavior. In addition, we wish to assess whether the range of values of x_0 and G is restricted within narrow limits, thereby determining the utility of applying the algorithm to particles whose shape is not known *a priori*.

Our semi-empirical theory has three free parameters: r, x_0, and G. The parameter r can be determined once the particle's shape is known. Alternatively, it can be found from comparison with the observed absolute scattered intensity for various size regimes. The optimum manner for evaluating x_0 is to compare Mie scattering calculations with the measured phase functions for randomly oriented particles having monodisperse size distributions and to find the smallest size at which significant differences occur. Such an approach is readily accommodated by analog measurements (*e.g.*, Zerull and Giese, 1974), but usually not by the direct visual ones (*e.g.*, Holland and Gagne, 1970). Alternatively, x_0 can be found by trial-and-error fits to measurements for polydispersed size distributions.

Finally, G is obtained by optimizing the fit of the theory to the measured phase function of polydispersed ensembles at intermediate and large angles of scatter.

Figures 3, 4, 6, and 7 compare our computed phase function with the observed function for a variety of particle shapes, as well as the predicted behavior of equal volume spheres and equal area spheres. Quite clearly, our algorithm provides a good fit to all these cases over the entire observed range of scattering angles. The joint influence of the small- and large-size regime is nicely illustrated in Figure 7. For this case, x_0 has a value close to the midpoint of the range of x values included in the size distribution function.

4. DISCUSSION AND CONCLUSIONS

The relatively simple structure of our theory provides insight into the similarities and differences in the scattering behavior of irregular particles and that of their equal volume spherical counterparts. First, polydispersed ensembles of *small* irregular particles,

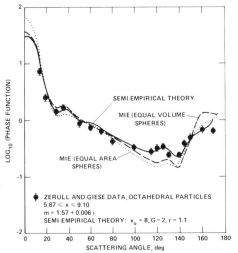

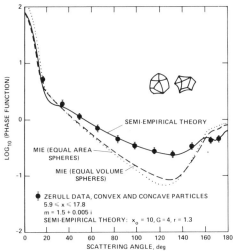

FIG. 6. Comparison of the measured phase function for scattering by an ensemble of octahedra, (dots with error bars; Zerull and Giese, 1974), with the predictions of Mie theory for equal volume and equal area spheres and with the semi-empirical theory of this paper.

FIG. 7. Comparison of the measured phase function for scattering by an ensemble of convex and concave particles (dots with error bars; Zerull, 1976), with the predictions of Mie theory for equal volume and equal area spheres and with the semi-empirical theory of this paper.

i.e., $x < x_0$ for all particles, are expected to exhibit essentially the same scalar scattering behavior as that of equivalent volume spheres. However, as discussed in Section 3.1, this agreement breaks down for very elongated particles having large $|m|$, with such particles having higher cross sections than their spherical counterparts.

Next, consider an ensemble of large irregular particles, *i.e.*, $x > x_0$ for all particles. According to equation (6a), the average *scattering* efficiency of such particles differs from that of equal volume spheres only by a factor equal to the ratio of their surface areas, r. This parameter is a weak function of particle shape. Therefore, the total extinction cross section and (unless $2m_i x > 1$) the single scattering albedo of the irregular particles is higher than that of the equivalent spheres.

The *shape* of the phase function for $x > x_0$ at *small* scattering angles is determined primarily by diffraction and external reflection. Therefore, it is not significantly affected by particle irregularities. Besides the elimination of resonance features such as the rainbow and glory, the main deviation in the scattering behavior of large, irregular particles from spheres arises in that component which is internally transmitted and refracted. These differences are especially important at intermediate and large angles of scatter. Usually, the sense of the deviation is a redistribution of energy from the forward hemisphere to the backward hemisphere. All the irregular particles for which we made comparisons, including the flat flakes of Holland and Gagne (1970), exhibit less forward scattering, as measured by $<\cos(\theta)> = \int p^*(\cos\theta)\, d\Omega/4\pi$, than their spherical counterparts (see Table 1).

The G and x_0 parameters of our theory can be related in a crude way to the shape of the irregular particles. Rays entering approximately equidimensional particles have a higher probability of experiencing total internal reflection than rays entering particles with one dimension significantly smaller than the other ones. Hence, G is expected to be smaller for equidimensional objects than for elongated ones (cf., for example, the G values of the cubes and flat flakes). For a fixed refractive index, x_0 is probably related to the roughness of the particles or more precisely to the presence or absence of large sharp corners. Rough particles should have small values of x_0. This supposition is consistent with the difference in the value of x_0 for cubes and octahedra shown in Table 1. There is also a suggestion that increasing the refractive index decreases x_0 (compare the second and third entries of Table 1). However, more data are needed to test this possibility.

TABLE 1

MODEL PARAMETERS DERIVED FROM LABORATORY MEASUREMENTS

Data[a]	m[b]	x_o[c]	G[c]	$<\cos \theta>$	$<\cos \theta>$ (equivalent) spheres)
Large cubes, x=5.9-17.8 $n(x)=n_o x^{-4}$	1.57 + 0.006i	4	1.5-2.0	0.540	0.672
All cubes, x=1.9-17.8 $n(x)=n_o x^{-2.5}$	1.57 + 0.006i	4	1.5-2.0	0.560	0.690
All cubes, x=1.9-17.8 $n(x)=n_o x^{-2.5}$	1.70 + 0.015i	2	1.5	0.482	0.681
Octahedra, x=5.9-9.1 $n(x)=n_o x^{-2.5}$	1.50 + 0.005i	7-8	2	0.520	0.570
Convex-concave, x=5.9-17.8 $n(x)=n_o x^{-2.5}$	1.50 + 0.005i	10	3-4	0.601	0.719
Flat plates or flakes x=2-20 log normal	1.57 + 0i	3	5	0.600	0.675

[a]The measurements on the cubes, octahedra, and convex-concave particles were made by Zerull and Giese (1974) and Zerull (1976), while those for the flat flakes were obtained by Holland and Gagne (1970). In the case of the convex-concave particles, we used an average of the separate measurements made on convex and concave particles.

[b]m is the complex index of refraction.

[c]x_o and G are parameters of the non-spherical particle scattering theory, which were derived by matching the measurements cited in the leftmost column. In the case of the cubes and the convex-concave particles, the parameter r was estimated to be about 1.3, while it was found to be about 1.1 for the octahedra. From strictly geometrical considerations, r should equal 1.3 for cubes. The parameters x_o, G, and r are defined in Section 4 of the text. The value of r for the flakes is not well-determined, as an equal-area comparison was initially used. Values of 1 < r < 2 are virtually indistinguishable for this case.

Question (REAGAN): Would you please define what you mean by surface area normalization for your semi-empirical theory?

Answer (CUZZI):   In our theory, we start with an ensemble of spheres of equal volume. For those spheres with size parameters less than x_0, we assume the total scattering is proportional to volume. For the spheres with $x > x_0$, we assume the total scattering is proportional to surface area, and therefore, increase the *scattering* but not absorption cross sections by a typical ratio (r) of surface area of the irregular particles to that of equal volume spheres.

REFERENCES

Atlas, D., Kerker, M. and Hirschfeld, W., 1953, *J. Atmos. Terrestrial Phys.*, *3*, 108-119.

Berry, C. R., 1962, *J. Opt. Soc. Amer.*, *52*, 888-895.

Greenberg, J. M., 1972, *J. Colloid Interface Sci.*, *39*, 513-519.

Greenberg, J. M., Pedersen, N. E., and Pedersen, J. C., 1961, *J. Appl. Phys.*, *32*, 233-242.

Greenberg, J. M., Wang, R. T., and Bangs, L., 1971, *Nature*, *230*, 110-112.

Hansen, J. E. and Travis, L. D., 1974, *Space Sci. Rev.*, *16*, 527-610.

Hodkinson, J. R., 1963, *in* "Electromagnetic Scattering," M. Kerker, Ed., Pergamon Press, p. 87-100.

Hodkinson, J. R. and Greenleaves, I., 1963, *J. Opt. Soc. Amer.*, *53*, 577-588.

Holland, A. C. and Gagne, G., 1970, *Appl. Opt.*, *9*, 1113-1121.

Jones, A. R., 1972, *J. Phys. D: Appl. Phys.*, *5*, L1-L4.

Pollack, J. B., and Cuzzi, J. N., 1979, submitted to *J.A.S.*

Proctor, T. D. and Barker, D., 1974, *Aerosol Sci.*, *5*, 91-99.

Proctor, T. D. and Harris, G. W., 1974, *Aerosol Sci.*, *5*, 81-90.

van de Hulst, H. D., 1957, "Light Scattering by Small Particles," John Wiley and Sons, New York.

Vouk, V., 1949, *Nature*, *162*, 330-331.

Zerull, R., 1976, *B. zur Phys. Atmopsh.*, *49*, 168-188.

Zerull, R. and Giese, R. H., 1974, *in* "Planets, Stars, and Nebulae Studies with Photopolarimetry," T. Gehrels, Ed., Univ. of Ariz. Press, p. 901-914.

SCATTERING AND ABSORPTION BY WAVELENGTH SIZE DISCS

Herschel Weil and C. M. Chu

Univ. of Michigan, Dept. of Electrical & Computer Engineering

Ann Arbor, Michigan 48109

ABSTRACT

An integral equation for the current density induced in flat lossy dielectric discs by incident polarized waves is solved by a partly analytic, partly numerical method designed specifically to be efficient for the disc shape with thickness small and radius comparable to the incident wavelength. The computed currents are then used to compute, for arbitrary direction and polarization of the incident wave, the bistatic scattering matrix elements, total scattering and absorption cross-sections. The extinction cross-section is found both by summing the latter two cross-sections and by using the optical theorem relating forward scatter and extinction.

1.   INTRODUCTION

The numerical procedure to solve the integral equation is a combination moment-finite element method tailored for thin circular discs. The general ideas involved together with details of their application for the special cases of the incident wave propagating in directions either parallel or perpendicular to the flat sides of the disc and with electric vector parallel to the flat sides are given in Chu and Weil (1976). Here we summarize the basic method, describe its application to the general case of arbitrary direction of incidence and state of polarization, and give sample numerical results.

2. THE INTEGRAL EQUATION AND ITS SOLUTION

The integral equation for current density J induced in a homogeneous dielectric object of refractive index n is

$$E^i = \frac{-1}{i\omega\epsilon_0}[\underline{L} \cdot J - \frac{J}{n^2 - 1}] \tag{1}$$

where $\underline{L}$ is the self-adjoint integral operator defined by

$$\underline{L} \cdot J = \iiint_D dv'J(R')G(R,R') - \iint_S ds'J(R') \cdot \hat{n}'(R')\nabla G(R,R') \tag{2}$$

D and S represent the volume and surface of the object and

$$G = \exp(i|R - R'|)/(4\pi|R - R'|). \tag{3}$$

All lengths are normalized by multiplication by $k_0 = 2\pi/\lambda_0$ where λ_0 is the free space wavelength. The time dependence of the fields is assumed to be of the form $e^{i\omega t}$.

Equation (1) is reduced to a set of linear algebraic equations via the method of moments. To do so $J(R)$ is expanded in a series of divergenceless basis functions, $W_j(R)$, so that

$$J(R) = -i\omega\epsilon_0 \sum_j \alpha_j W_j(R). \tag{4}$$

The basis functions should satisfy the requirements $\nabla \cdot W_i = 0$ in D and $W \cdot \hat{n} \neq 0$ on S to ensure the similar properties (which exist in a homogeneous medium) for $J(R)$ itself.

Substitution of (4) into (1) yields

$$E^i = \sum_i \alpha_i [\underline{L} \cdot W_i - (1/(n^2 - 1))W_i]. \tag{5}$$

Using the inner product

$$<f,g> = \iiint dv' \; f(R') \cdot g(R') \tag{6}$$

with equation (4) then yields an algebraic matrix equation to be solved for the α_i,

$$\langle \mathbf{W}_j, \mathbf{E}^i \rangle = \sum_i \alpha_i [Z_{ij} - W_{ij}/(n^2 - 1)] \equiv \sum_i \alpha_i z_{ij} \qquad (7)$$

where

$$Z_{ij} = \langle \mathbf{W}_i, \underline{\mathbf{L}} \cdot \mathbf{W}_j \rangle, \quad W_{ij} = \langle \mathbf{W}_i, \mathbf{W}_j \rangle \qquad (8)$$

The symmetric matrix $[Z_{ij} - W_{ij}/(n^2 - 1)]$ is termed an impedance matrix. All material effects are confined to the factor $(n^2 - 1)^{-1}$ while shape effects are included in Z_{ij} and W_{ij}.

To specify the basis functions $\mathbf{W}_i$ we use a cylindrical coordinate system (ρ, ϕ, z) with z along the symmetry axis of the disc. The disc fills a region given by $-\delta/2 \leq z \leq \delta/2$, $0 \leq \rho \leq a$. Then we choose the $\mathbf{W}_i$ from the following sets of functions: transverse electric modes

$$\mathbf{W}_E^{(n,m)} = \nabla \times \{\hat{z} R(\rho) Z_n(z) \phi_m(\phi)\} \qquad \text{(TE mode)} \qquad (9)$$

and transverse magnetic modes

$$\mathbf{W}_M^{(n,m)} = \nabla \times \mathbf{W}_E^{(n,m)} \qquad \text{(TM mode)} \qquad (10)$$

where

$$Z_n(z) = z^n; \quad \phi_m(\phi) = \sin m\phi \quad \text{or} \quad \phi_m(\phi) = -\cos m\phi . \qquad (11)$$

The form of $Z_n(z)$ is chosen to permit easy integration over z when evaluating the impedance expressions. In accord with the thin disc assumption, only terms of order δ^2 or greater will be kept in the impedance, hence only n=0 and 1 functions will be needed. The following odd TE modes, resulting from the choice of sin mϕ in (11), are obtained:

$$W_E^{(0,m)} = \hat{\rho}(mR(\rho)/\rho)\sin m\phi + \hat{\phi}R'(\rho)\cos m\phi \qquad \text{(TE0 ODD)}$$

$$W_E^{(1,m)} = \hat{\rho}(mR(\rho)/\rho)z \sin m\phi + \hat{\phi}R'(\rho)z \cos m\phi, \qquad \text{(TE1 ODD)} \tag{12}$$

The corresponding TM modes contain $\hat{z}$ as well as $\hat{\rho}$ and $\hat{\phi}$ components. For even modes m is to be replaced by $m\phi + (\pi/2)$ in (11) and (12). A combination of TE and TM mode **W** functions will be required.

With these **W** functions there is no coupling between the different values of m; that is, the impedance terms corresponding to the use of $W^{(n,m)}$ and $W^{(n,m')}$ are zero when $m \neq m'$. Likewise there is no coupling between even and odd modes. Further decoupling is achieved by dropping all terms of order δ^3 or higher when carrying out the volume integrations called for in determining the matrix elements. As a result the impedance matrix has many zero elements and can be partitioned as shown in Figure 1 which greatly facilitates the numerical solution. The work of Chu and Weil (1976) was limited to special cases where only TE modes with n=0 were directly excited, and hence only matrix terms involving these modes were taken to be not zero [the z(TE0,TE0) terms] of Figure 1.

A set $\{R_i(\rho)\}$ of radial functions was chosen so as to facilitate the integration. Each $R_i(\rho)$ is zero over all but a fractional part

$$
\underset{\approx}{z} =
\begin{bmatrix}
\underset{\approx}{z}(m=1, \text{ODD}) & 0 & 0 & \cdots \\
0 & \underset{\approx}{z}(m=1, \text{EVEN}) & 0 & \cdots \\
0 & 0 & \underset{\approx}{z}(m=2, \text{ODD}) & \cdots \\
\vdots & \vdots & \vdots & \ddots
\end{bmatrix}
$$

$$
\text{and,} \quad \underset{\approx}{z}(m) =
\begin{bmatrix}
z(\text{TMo,TMo}) & & 0 \\
& & z(\text{TEo,TEo}) \quad z(\text{TEo,TMi}) \\
0 & & z(\text{TEo,TMi}) \quad z(\text{TMi,TMi})
\end{bmatrix}
$$

FIG. 1.   The impedance matrix: $z_{ij} = Z_{ij} - W_{ij}(n^2 - 1)$.

of the interval $0 \leq \rho \leq a$, and in the subinterval where it is not zero R_i is a spline function consisting of segments of second order polynomials. The R_i are illustrated in Figure 2 which also shows how the subintervals overlap. The subintervals are taken small enough to permit the integrals to be evaluated by approximate means with adequate accuracy and yet avoid an excessively large number of intervals.

The integration over the singularity region in the neighborhood of $|R-R'|=0$ is then evaluated to the desired accuracy by reducing the problem to the singularity integration for a statics problem plus correction terms, all of which are evaluated analytically.

The W_{ij} and excitation terms $\langle W_j \cdot E^i \rangle$ each involve only a single volume integral and can be evaluated analytically with our choice of polynomial functions for the $R_i(\rho)$ by straightforward integration.

The Z_{ij} each involve double volume and surface integrals with singular integrands. They are carried out by partly numerical and partly analytical methods in which the two z integrations are done analytically to $O(\delta^2)$. The two angular integrations are reduced analytically to a single integral which is approximated by a finite sum. The analytic treatment of the singularities plus the use of average values for the ρ dependent functions averaged over the subintervals reduce the radial integrals to finite sums. This somewhat lengthy analysis leads to formulas which were programmed and used for computation of the α_i. The formulas and a description of the computational code are in Weil and Chu (1978).

3.   SCATTERING AND ABSORPTION COMPUTATION

Once α_i and hence J (in equation (4)) are determined, classical EM theory permits the reradiated and the absorbed fields and power

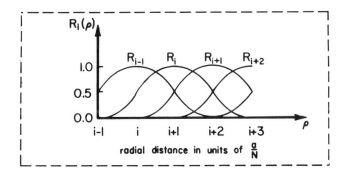

FIG. 2.   The radial functions.

to be obtained by straightforward integrations over the disc volume.
In particular, the reradiated fields are given by application of the
vector Huygens' principle while the absorbed power is obtained by
integrating the Joule heating throughout the volume. The alternative
use of the Optical Theorem for the extinction gives a means of
checking the accuracy of the computations; for example, whether
"enough" terms were retained in equation (4). The integrations were
carried out analytically, leading to sums for the various desired
parameters such as bistatic cross section σ_B, absorption cross section
σ_A, total or integrated scattering cross section σ_{TS}, etc.

As illustrations of the results we give data computed for discs
of circumference/wavelength = 2.95 and refractive index $n=1.13-i \times$
(0.2273). The discs are irradiated by plane polarized plane waves at
angles of incidence (θ_0) of 0° (broadside), 45° and 90° (edge-on).
As illustrated in Figure 3 the waves are considered to be incident in
the xz plane $(\phi_0=0)$, each with y directed polarization. The data is
given in terms of normalized cross sections, $k_0^2\sigma$. For given incident
directions (θ_0,ϕ_0) the bistatic data are given as functions of
scattering angles θ_S,ϕ_S measured with respect to the fixed coordinate
system (not the incoming direction). The bistatic data is shown
graphically in Figure 4 for variable θ in the xy plane, x>0, (azimuthal angle $\phi=0$) to illustrate forward scattering, the yz plane
$(\phi=90°)$ and the xz plane, x<0$(\phi=180°)$ to illustrate backscattering.

These data clearly indicate the effects on both scattering and
absorption of the imaginary part of the refractive index. They also
show clearly that when the thin disc is irradiated broadside it
scatters like a Rayleigh particle in that the scattering pattern is

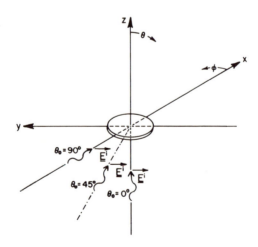

FIG. 3.   Disc and incoming wave orientations.

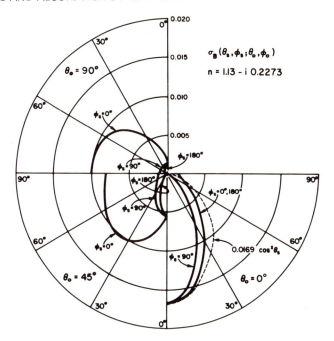

FIG. 4.   Angular scattering patterns (bistatic differential cross sections) for a disc of radius 2.95.   Lengths are in units of k^{-1}. Each quadrant shows curves for a different one of the incoming waves illustrated in Fig. 3.   Each curve is for scattered power in a fixed azimuthal plane and varying polar angle θ.

CROSS – SECTIONS   (dB)

"upper" numbers: n = 1.13 – i 0.2273
"lower" numbers: n = 1.13 – i 0.0

$\theta_o = 0°$	$\theta_o = 45°$	$\theta_o = 90°$
	ABSORPTION	
1.61	0.55	0.55
-∞	-∞	-∞
	TOTAL SCATTERING	
-13.4	-14.8	-16.4
-19.1	-20.6	-22.1
	EXTINCTION (FROM ABS. AND TOT. SCAT.)	
1.74	0.67	0.65
-19.1	-20.6	-22.1
	EXTINCTION (FROM FORWARD SCAT.)	
1.80	0.67	0.65
-17.4	-21.4	-22.9

FIG. 5.   Cross sections for disc of radius 2.95 (lengths are in units of k^{-1}).

asymmetric in the forward and backward hemispheres although there is, nonetheless, a clear deviation from the $\cos^2\theta$ Rayleigh pattern. When the disc is irradiated at 45° or edge-on there is much stronger radiation in the forward than in the backward hemisphere. This is to be expected since there is considerable phase shift across the disc diameter.

The corresponding integrated data, total scattering absorption and extinction are tabulated in Figure 5.

Acknowledgement. This work was supported by the National Oceanic and Atmospheric Administration under Grant 04-78-B01-1.

Question (SCHAEFER): Have you compared any of your disc results to Asano's oblate spheroids?

Answer (WEIL):   Not yet.

Question (CUZZI):   Did you obtain any comparisons of forward scattering efficiencies as compared to those of spheres?

Answer (WEIL):   No, but it would be interesting to do so.

REFERENCES

Chu, C. M. and Weil, H., 1976, *J. Computational Phys.* <u>22</u>, 111-124.
Weil, H. and Chu, C. M., 1978, *Scattering and Absorption of Electro-magnetic Radiation by Clouds of Disc Shaped Aerosols*, Univ. of Michigan Report 016234-1-F, Ann Arbor, Michigan, December 1978.

PERTURBATION APPROACH TO LIGHT SCATTERING BY

NON-SPHERICAL PARTICLES

J. T. Kiehl
State Univ. of N.Y. at Albany, Department of Atmospheric
Science, Albany, NY 12222, and National Center
for Atmospheric Research, Boulder, CO 80307

M. W. Ko
Atmospheric and Environmental Research, Cambridge, MA 12138

A. Mugnai
CNR-LPS, Elettrofisica Atmosferica, Frascati, Italy

Petr Chýlek
State Univ. of N.Y. at Albany, Atmospheric Sciences
Research Center, Albany, NY 12222, and National
Center for Atmospheric Research, Boulder, CO 80307

ABSTRACT

Applying perturbation theory we have derived the first order
perturbation corrections to the scattering characteristics for the
case of light scattering by slightly deformed spheres. Numerical
results are compared with those obtained using the extended boundary
condition method. The range of applicability of the first order
perturbation corrections are discussed.

The application of perturbation theory to light scattering by
non-spherical particles has experienced very little development since
it was first derived by Yeh (1964, 1965) and Erma (1968). It is the
purpose of this report to investigate the application of first order
perturbation theory to light scattering by various non-spherical
particles. The results of perturbation theory are then compared with
calculations for the same particles using the extended boundary condition method (EBCM) developed by Waterman (1971), and Barber and
Yeh (1975).

135

A general formalism for calculating the nth order perturbation term has been developed by Yeh (1964, 1965) and Erma (1968). The particle is assumed to be of the form:

$$r = r_s \left(1 + \varepsilon\, f(\theta,\phi)\right) \qquad (1)$$

where r_s is the radius of an unperturbed sphere, $f(\theta,\phi)$ describes the shape of the irregularity on the sphere. It is further assumed that $\varepsilon \ll 1$ and that $\left|\varepsilon f(\theta,\phi)\right| < 1$. The calculations reported herein used the following functional forms for $f(\theta,\phi)$:

$$f(\theta,\phi) = T_2(\cos\theta) \qquad (2)$$

$$f(\theta,\phi) = T_4(\cos\theta) \qquad (3)$$

where $T_n(\cos\theta)$ is the nth order Chebyshev polynomial.

The shapes of these particles are illustrated in Figure 1. The particle is assumed to be oriented such that it is rotationally symmetric about the z-axis. The incident light is assumed to be traveling in the z-direction with E vector in the x-direction. First order perturbation theory assumes that the scattering coefficients for the non-spherical particle can be expressed as

$$a_n = a_n^0 + \varepsilon\, a_n^1 , \qquad (4)$$

$$b_n = b_n^0 + \varepsilon\, b_n^1 , \qquad (5)$$

where a_n^0, b_n^0 are the Mie scattering coefficients for a sphere of radius r_s; a_n^1, b_n^1 are the first order corrections due to the non-sphericity of the particle.

Calculations were performed for the case where the index of refraction $m = 1.5$ and $\varepsilon = \pm 0.1$. Results for other indexes of refraction and values of ε have been reported in Chýlek $et\ al.$ (1978). The efficiencies are calculated by normalizing the extinction cross sections by πr_s^2.

Figure 2 shows the extinction efficiency for both perturbation theory and EBCM for the case of $f(\theta,\phi) = T_2(\theta)$ and $\varepsilon = +0.1$. The extinction efficiency for a sphere of radius r_s is also shown in Figure 2. It is evident from Figure 2 that perturbation theory agrees quite well with EBCM for $x_s < 6$. For values of $x_s > 6$ there

(a) $T_2(\cos\theta)$ (b) $T_4(\cos\theta)$

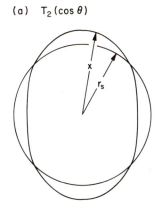   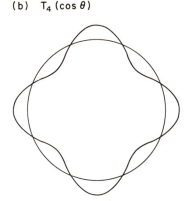

FIG. 1.   The two shapes considered are: a) $r = r_s\left(1 \pm \epsilon\, T_2(\cos\theta)\right)$, b) $r = r_s\left(1 \pm \epsilon\, T_4(\cos\theta)\right)$.

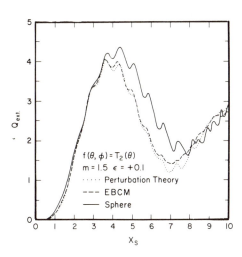

FIG. 2.   The extinction efficiency for light scattering by a particle whose shape is given by $r = r_s\left(1 + 0.1\,T_2(\cos\theta)\right)$. The solid line (——) is for a sphere with $r = r_s$. The dashed line (---) represents the exact results of the EBCM. The dotted line (···) represents the results of first order perturbation theory.

is very poor agreement between perturbation theory and EBCM due to the narrowness of the resonance in the extinction efficiency. The effect of the width of a resonance on perturbation theory has been discussed by Chýlek $et\ al.$ (1978). There it was shown that in the resonance region one would expect perturbation theory to apply when

$$|\epsilon| < \Gamma/2\ x_s\ , \qquad (6)$$

where Γ is the half width of a Lorentzian shaped resonance. Equation (6) implies that for large values of x_s, where Γ is expected to be small, one would need a very small ϵ for perturbation theory to agree with the exact calculations.

Figure 3 shows the extinction efficiency for the same shape function but with $\epsilon = -0.1$. The observations made above for Figure 2 apply to this case.

The difference between perturbation theory and EBCM can best be illustrated by considering the relative difference between the results obtained using these two methods. This difference is defined as

$$\delta = \frac{Q_{ext}^{Pert} - Q_{ext}^{EBCM}}{Q_{ext}^{EBCM}} \tag{7}$$

which is a function of x_s. Figure 4 shows the absolute value of equation (7) for the case of $\varepsilon = +0.1$ and $f(\theta,\phi) = T_2(\theta)$. Since terms of order ε^2 have been neglected in these calculations an error of 10^{-2} would be expected. It can be seen from Figure 4 that the difference lies between 10^{-3} and 10^{-1}. A striking feature of this

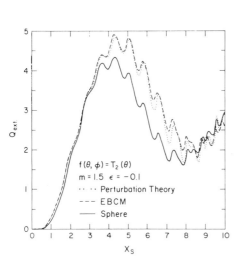

FIG. 3. The extinction efficiency for light scattering by a particle whose shape is given by $r = r_s\left(1 - 0.1\,T_2(\cos\theta)\right)$. The solid line (———) is for a sphere of radius $r = r_s$. The dashed line (---) represents the exact results of the EBCM. The dotted line ($\cdots$) represents the results of first order perturbation theory.

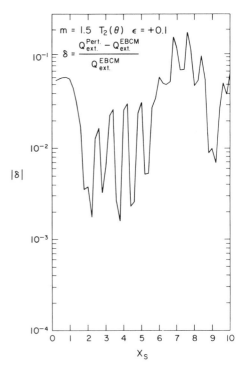

FIG. 4. The absolute value of the relative difference between perturbation theory and the EBCM for the particle: $r = r_s\left(1 + 0.1\,T_2(\cos\theta)\right)$ as a function of $x_s = 2\pi r_s/\lambda$.

figure is the oscillatory behavior of $|\varepsilon|$. These oscillations are
due to the resonance effect discussed above for Figures 2 and 3.
There exists a one-to-one correspondence between the position of the
maxima in relative difference in Figure 4 and the position of the
resonance of the extinction curve. An overall growth in absolute
relative difference can be observed for increasing values of x_s due
to the narrowing of the resonances.

Similar calculations for $f(\theta,\phi) = T_4(\theta)$ and $\varepsilon = +0.1$ are shown
in Figure 5. For this case little difference can be seen to exist
for $x_s < 7$. It is also observed that the $T_4(\theta)$ particle has a
smaller extinction than the sphere for this size range. Figure 6
illustrates the extinction efficiency for a shape function $f(\theta,\phi) = T_4(\theta)$ and $\varepsilon = -0.1$.

Figure 7 shows the absolute value of the relative difference for
the $T_4(\theta)$ particle with $\varepsilon = +0.1$ as a function of x_s. The relative
difference from this figure lies between 2.5×10^{-4} to 2.2×10^{-1}.
The oscillatory behavior of this difference correlates with the resonances in Figure 5.

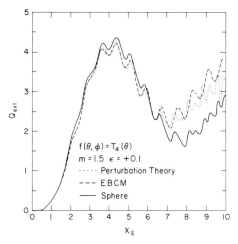

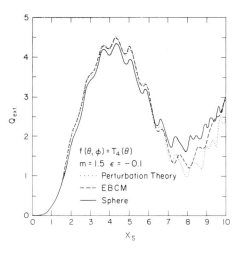

FIG. 5.   The extinction efficiency for light scattering by
a particle whose shape is given
by $r = r_s\left(1 + 0.1\ T_4(\cos\theta)\right)$.
The solid line (——) is for a
sphere of radius $r = r_s$. The
dashed line (- - -) represents
the exact results of the EBCM.
The dotted line ($\cdots$) represents
the results of first order perturbation theory.

FIG. 6.   The extinction efficiency
for light scattering by a particle
whose shape is given by $r = r_s$
$(1 - 0.1\ T_4(\cos\theta))$. The solid
line (——) is for a sphere of radius
$r = r_s$. The dashed line (- - -)
represents the exact results of
the EBCM. The dotted line ($\cdots$)
represents the results of first
order perturbation theory.

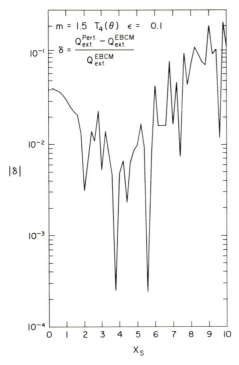

In conclusion, this study indicates that first order perturbation theory as applied to light scattering by non-spherical particles is limited to small deformations and low values of the size parameter. The limitation arises because of the breakdown of first order perturbation theory in the region of resonances of the scattering coefficients. However, these problems can be eliminated by taking higher order terms in the perturbation expansion. The computational feasibility of such calculations is possible since these calculations require little storage and computer time.

FIG. 7.   The absolute value of the relative difference between perturbation theory and the EBCM for the particle: $r = r_s\left(1 + 0.1 \; T_4(\cos\theta)\right)$ as a function of $x_s = 2\pi r_s/\lambda$.

Acknowledgement.   This work was supported in part by the U.S. Army Research Office.   The National Center for Atmospheric Research is sponsored by the National Science Foundation.

REFERENCES

Barber, P. and Yeh, C., 1975, *Appl. Opt.*, *14*, 2864.
Chýlek, P., Kiehl, J. T., Ko, M. W., 1978, Third Conference on
    Atmospheric Radiation, American Meteorological Society, 1978.
Erma, V., 1968, *Phys. Rev.*, *179*, 1238.
Waterman, P. C., 1971, *Phys. Rev. D*, *3*, 825.
Yeh, C., 1964, *Phys. Rev.*, *135*, A1193.
Yeh, C., 1965, *J. Math. Phys.*, *6*, 2008.

EXACT CALCULATIONS OF SCATTERING FROM MODERATELY-NONSPHERICAL T_n-PARTICLES: COMPARISONS WITH EQUIVALENT SPHERES

Warren Wiscombe and Alberto Mugnai

National Center for Atmospheric Research

Boulder, Colorado 80307

ABSTRACT

We study extinction efficiency, absorption efficiency, and forward and backscatter efficiency for rotationally symmetric particles of the form $r(\theta) = r_o[1 + \varepsilon T_n(\cos \theta)]$. Using the Extended Boundary Condition Method, as embodied in a computer code from Dr. Peter Barber, we have compared the various efficiencies, for T_2- and T_4-particles and $0 < \varepsilon \leq 0.2$, with the corresponding results for equal-volume and equal-projected-area spheres. A range of 0 to 10 in equivalent Mie size parameter x was considered, and results are given for nose-on incidence, for random orientation, and for random orientation plus a running mean over a size interval $\Delta x = 1$.

1. INTRODUCTION

There have been a number of measurements of scattering from non-spherical particles, both of micron-sized particles in the optical range and of centimeter-sized particles in the microwave range. While a few of these measurements have been contradictory, the majority indicate certain common features. Mie theory for equal-volume (E.V.) or equal-projected-area (E.P.A.) spheres gives a reasonable approximation for the extinction and absorption efficiencies and for the phase function out to 40°-60° or so, provided all but the main features in Mie results are averaged or smoothed out in some way; and backscattering is often considerably different from what Mie theory would predict. (Other features, such as the enhancement of side-scattering, are not of direct concern in this paper.)

At the same time, several semi-empirical methods have been put forward to explain non-spherical particle scattering which, if correct, would be of great value.

In order both to understand the measurements and to test the semi-empirical theories, exact calculations of non-spherical particle scattering are desirable. The Extended Boundary Condition Method (EBCM) of Waterman (1965), as improved by Barber and Yeh (1975), is capable of providing such calculations. Professor Barber generously furnished us with his EBCM computer code, which we have developed somewhat further in order to make it more efficient.

We have used the EBCM code to try to answer questions about the features of non-spherical scattering which are important for flux calculations in radiative transfer and for LIDAR measurements. In particular:

1) How accurately are the extinction, absorption, scattering, forward-scattering and back-scattering efficiencies of non-spherical particles fitted by equivalent spheres?
2) How do orientation averaging and/or size averaging improve the fit?
3) Does surface roughness or concavity have a systematic effect?
4) As a particle departs in a continuous manner from a sphere, how do the efficiencies in question (1) behave?

Below we present a sample of results from this study.

2.   THE PARTICLE SHAPE

We devoted considerable thought to the selection of a particle shape. (The EBCM can handle almost any shape, although great elongation or deep concavities can result in ill-conditioning or prohibitive computation times.) We wanted a shape with a single parameter describing "non-sphericity," defined as degree of departure from a sphere and a second parameter describing surface roughness (size is a third parameter not essentially part of the shape). The shape also had to look somewhat realistic, be somewhat symmetrical, and be "general" in the sense that a wide variety of particles might reasonably be approximated by it. A final consideration was that the EBCM calculations would become prohibitive for anything but a rotationally symmetric shape.

With all these considerations in mind, we hit upon the following shape:

$$r = r_o[1 + \varepsilon T_n(\cos \theta)] \tag{1}$$

where r and θ are the usual spherical coordinates of the particle's surface, $T_n(\cos \theta) = \cos n\theta$ is the Chebyshev polynomial of degree n, ε is the non-sphericity or deformation parameter, and n is the surface roughness parameter (equal to the number of peaks or valleys, which are all of equal amplitude and uniformly spaced in angle around the particle). All the T_n-particles except T_1 develop concavity as $|\varepsilon|$ increases, starting at $|\varepsilon|=1/17$ for n = 4. For $|\varepsilon|<0.2$ the T_2-particles resemble spheroids, and for $\varepsilon = 1/17$ the T_4-particles resemble right circular cylinders with rounded edges.

Since the T_n form a complete set, any reasonable axisymmetric particle can be approximated by

$$r = r_o[1 + \sum_{n=1}^{N} \varepsilon_n T_n(\cos \theta)] \quad . \tag{2}$$

Furthermore, by the well-known principle of "economization of series," equation (2) gives a *better* minimax approximation, for a given number N of terms, than any other polynomial expansion. Thus, it is clearly of value to study scattering by elementary T_n-particles, even though the scattering properties of equation (2) cannot be obtained just by adding the same properties for the elementary T_n-particles composing that shape (unless $\varepsilon_n \ll 1$).

Our initial investigation has only been for T_2- and T_4-particles and for $-0.2 \leq \varepsilon \leq 0.2$. T_2- and T_4-particles with $|\varepsilon|$ = 0.1 and 0.2 are shown in Figure 1 where the sign of ε is denoted by a superscript $\pm$; *e.g.*, $T_2^{\pm}$. For larger values of n or ε, the EBCM calculations were difficult to converge, undoubtedly because of the increasing concavity; and because of computer time limitations, we restricted our study to size parameters from 0 to 10. (For our E.V. plots, size parameter is defined as $x = (6\pi^2 V/\lambda)^{1/3}$, where V is particle volume and λ is wavelength.)

3. EXTINCTION, ABSORPTION, AND SCATTERING EFFICIENCY

Figure 2 shows extinction and absorption efficiencies Q_{ext} and Q_{abs} vs. size parameter for $T_2^{\pm}$ particles with $|\varepsilon|$ = 0.1, compared to the same efficiencies for E.V. and E.P.A. spheres; transparent, moderately absorbing, and highly absorbing cases are shown (indices of refraction 1.5, 1.5-0.05i, and 1.5-i, respectively). The radiation is incident along the axis of rotation indicated in Figure 1 ("nose-on"). Clearly, for size parameters x < 4 the E.V. assumption is much better, while by x = 10 it is already clear that the E.P.A. assumption is better for Q_{abs} in both absorbing cases, and for Q_{ext} in the highly absorbing case. Presumably, the E.P.A. assumption would also be better for larger transparent particles, since there are theoretical grounds for believing this to be so, but this is not yet

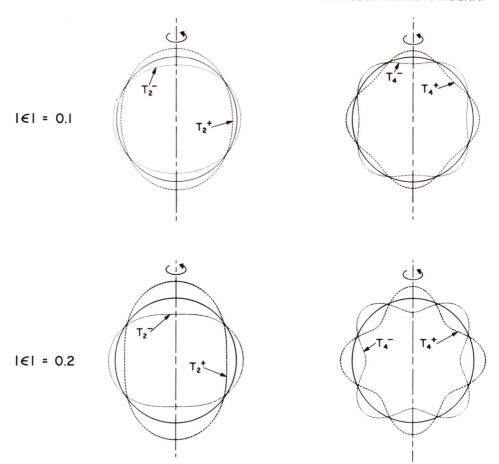

FIG. 1. $T_2^{\pm}$ and $T_4^{\pm}$ particle shapes with $|\epsilon|$ = 0.1 and $|\epsilon|$ = 0.2.

manifest by x = 10. An obvious conclusion is that, at least for absorbing particles, the best approximating sphere has E.V. for smaller sizes, transiting in some way to E.P.A. for larger sizes.

Figure 3 shows Q_{ext} and Q_{abs} for $|\epsilon|$ = 0.1 particles at nose-on incidence with $T_2^{\pm}$ results in the upper row and $T_4^{\pm}$ results in the lower row; E.V. spherical results are shown for comparison. The $T_4^{\pm}$ extinction efficiencies hew more closely to the E.V. spherical results than do the $T_2^{\pm}$ efficiencies out to x ∿ 7, probably because the surface roughness on $T_4^{\pm}$ particles is on a finer scale (compared to the wavelength), making them appear more spherical to the incoming radiation. Even beyond x = 7 the T_4^{-} results stay remarkably close to the spherical ones while the T_4^{+} results deviate considerably more.

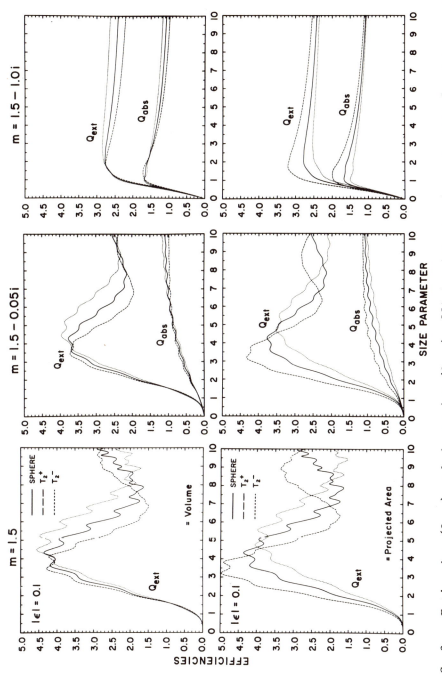

FIG. 2. Extinction (Q_{ext}) and absorption (Q_{abs}) efficiencies vs. size parameter for spheres and, for $T_2^{\pm}$ particles at nose-on incidence, vs. equal-volume size parameter in upper row and equal-projected-area size parameter in lower row. Results are shown for three refractive indexes (m).

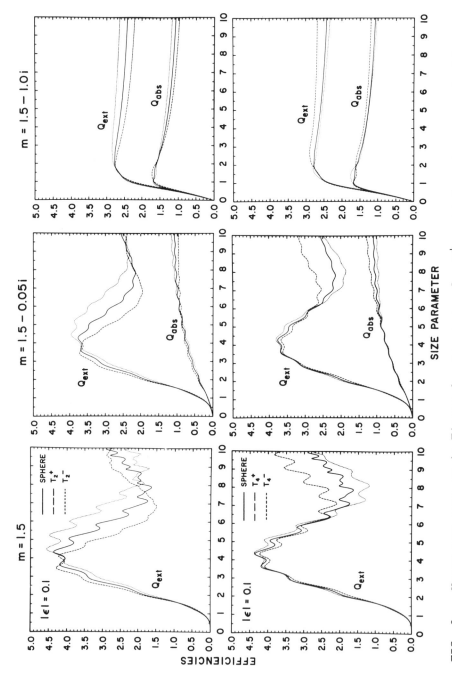

FIG. 3.   Upper row same as in Figure 2.   Lower row for $T_4^\pm$ particles at nose-on incidence.

In Figure 4, we have orientation-averaged (the symbol < >) the efficiencies of Figure 3. This eliminates, or at least greatly damps, the oscillatory structure of the $T_2^{\pm}$ and $T_4^{\pm}$ curves and brings them into much closer agreement with the E.V. spherical curves, as we would expect. Indeed, to plotting accuracy the spherical and non-spherical curves cannot be distinguished in some instances.

A feature which is evident in Figure 4, as it was in Figures 2-3, is that the spherical approximation is considerably better for the absorption efficiency than for the extinction or scattering efficiencies. While it would be premature to enunciate this as a general principle, our own results, including many not shown, are certainly compelling in this regard.

Another interesting observation is that concavity, present in the T_4 but not the T_2 cases, tends to systematically elevate the efficiencies for the larger particles ($x > 7$), a result which is in concord with the microwave-analogue measurements of Zerull (1976). This is not evident in the transparent case, but becomes so when a size-averaging is performed to smooth out the many small oscillations.

4. BACKSCATTER EFFICIENCY

The backscatter efficiency (which would be unity for isotropic scattering) is of great interest to the radar meteorologist and to users of LIDAR.

Figure 5 shows the orientation-averaged backscatter efficiency for $T_2^{\pm}$ and $T_4^{\pm}$ particles, compared with that for E.V. spheres. The spherical and non-spherical results are quite close out to $x \sim 4$ or so, and indeed all the way out to $x = 10$ (and farther, presumably) for the highly absorbing case. Elsewhere, the differences are large, due primarily to the large oscillations in the spherical results.

In practice, one will usually be dealing with a range of particle sizes. To see what effect this will have, we have made a running mean of the results in Figure 5, using a size parameter interval $\Lambda x = 1$. These size-averaged backscatter efficiencies are shown in Figure 6. Although this averaging is not sufficient to entirely eliminate the oscillations, it is sufficient to show that for $x > 3$ E.V. spheres usually overestimate, but also occasionally underestimate, the non-spherical backscatter; the differences depend in a complicated way on the refractive index, shape, and size of the particles. Highly absorbing particles are the exception; they backscatter very much like spheres over the full range of shapes and sizes we considered.

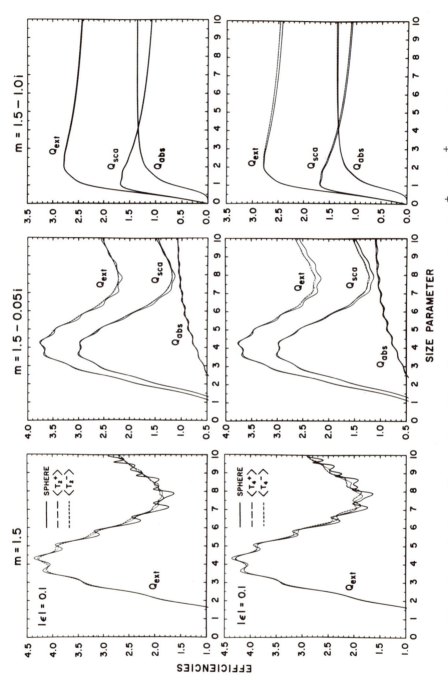

FIG. 4.   Same as Figure 3, except for *randomly oriented* $T_2^\pm$ and $T_4^\pm$ particles.

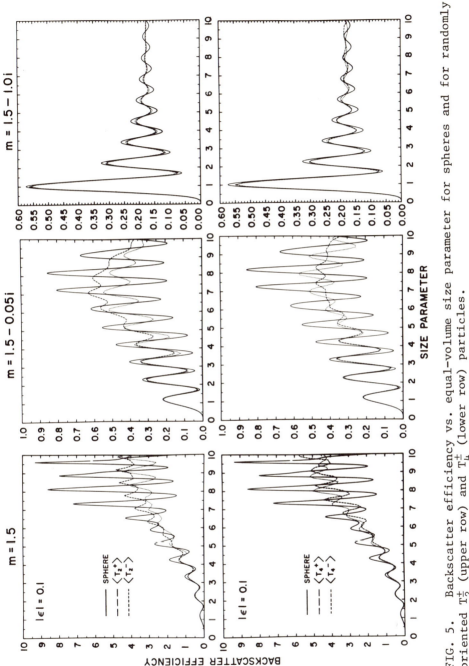

FIG. 5. Backscatter efficiency vs. equal-volume size parameter for spheres and for randomly oriented $T_2^\pm$ (upper row) and $T_4^\pm$ (lower row) particles.

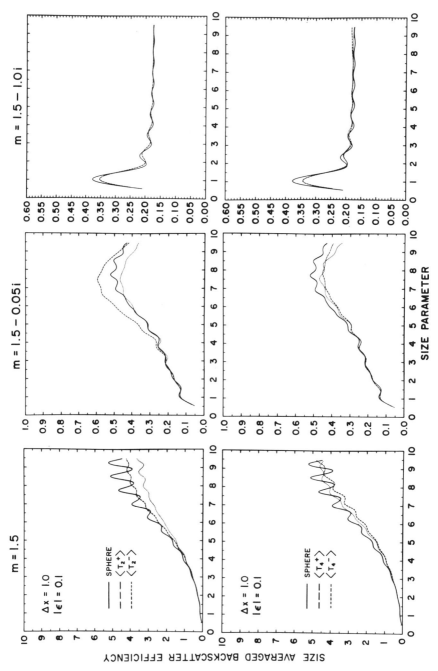

FIG. 6. Same as Figure 5, except a running mean over an interval of width Δx = 1 has been performed on each curve (x = size parameter).

5. FORWARD SCATTERING EFFICIENCY

Forward scattering efficiency is defined in a manner analogous to backscatter efficiency. Its values tend to be much larger, however, due to the well-known enhancement of forward scattering by diffraction.

The curves for forward scattering efficiency bear a great resemblance to those for extinction efficiency, which is not surprising in view of the optical theorem. Therefore, we show only orientation- and size-averaged values in Figure 7, with the T_n^+ and T_n^- values averaged together to boot.

Out to $x = 4$ the non-spherical and E.V. spherical results are in very close agreement. From $x = 4$ to $x = 9$, however, the non-spherical particle forward scattering is distinctly *larger*, more so for the concave T_4 particles than for the convex T_2-particles. The trend near $x = 9$ is such that T_4-particles will probably continue to forward scatter more than E.V. spheres out to larger sizes. T_2-particles, on the other hand, have already begun to forward scatter *less* than E.V. spheres at $x = 9$ for the transparent and moderately absorbing cases, and this trend also seems likely to continue to larger sizes. Only for the highly absorbing case do both T_2 and T_4-particles always scatter more in the forward direction than E.V. spheres.

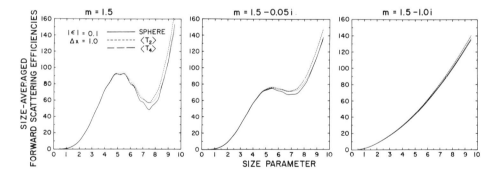

FIG. 7.   Forward scattering efficiencies averaged over $\Delta x = 1$ vs. equal-volume size parameter for spheres and for randomly oriented T_2 and T_4 particles.

Question (GOEDECKE): How did you weight your size averaging for equivalent spheres? It is possible that a match according to projected area, weighted with what it would be as you rotate the nonspherical particle through equal angular increments with respect to the incident beam, might be more reasonable than a flat weighting over some Δx.

Answer (WISCOMBE):   We indeed used a flat weighting over a size parameter interval Δx = 1. Our purpose was to average out the resonance structure in the curves without losing the major oscillations. For this purpose, we did not consider the actual form of the size distribution to be very important, but your suggestion sounds like a good one (although the actual calculation of projected area in arbitrary orientation is quite difficult, as you will find out if you try to do it).

REFERENCES

Barber, P. W. and Yeh, C., 1975, *Appl. Opt.* _14_, 2864-2872.
Waterman, P.C., 1965, *Proc. IEEE* _53_, 805-812.
Zerull, R. H., 1976, *Beitr. Phys. Atmos.* _49_, 166-188.

SURFACE WAVES IN LIGHT SCATTERING BY

SPHERICAL AND NON-SPHERICAL PARTICLES

Petr Chýlek
State Univ. of N.Y. at Albany, Atmospheric Sciences
Research Center, Albany, NY 12222, and National
Center for Atmospheric Research, Boulder, CO 80307

J. T. Kiehl
State Univ. of N.Y. at Albany, Department of Atmospheric
Science, Albany, NY 12222, and National Center
for Atmospheric Research, Boulder, CO 80307

M. K. W. Ko
Atmospheric and Environmental Research, Cambridge, MA 02138

A. Ashkin
Bell Telephone Laboratories, Holmdel, NJ 07733

ABSTRACT

Resonances in partial wave scattering amplitudes a_n and b_n are responsible for the ripple structure of the extinction curve and for sharp peaks in the backscattering (glory). A connection between the resonances and surface waves is suggested.

1. RESONANCES IN PARTIAL WAVE AMPLITUDES a_n AND b_n

It is convenient to write the Mie partial wave amplitudes a_n and b_n in the form

$$a_n = \frac{A_n}{A_n - iC_n} \quad (1a) \; ; \quad b_n = \frac{B_n}{B_n - iD_n} \quad (1b)$$

where A_n, C_n, B_n and D_n are real functions for real refractive index m. Their explicit form has been given elsewhere (Chýlek, 1973).

153

For simplicity of discussion we consider a real index of refraction
m. Then it follows that the $Re\{a_n\}$ and $Re\{b_n\}$ reach a series of
absolute maxima $Re\{a_n\} = 1$ and/or $Re\{b_n\} = 1$ at such values of
x ($x = 2\pi r/\lambda$ is a size parameter) where $Im\{a_n\} = 0$ and/or $Im\{b_n\} = 0$.
At sufficiently high values of n, the first maximum in each partial
wave amplitude a_n and b_n has the form of a sharp peak. In Figures
1A and 1B the first peaks in $Re\{a_{10}\}$ and $Re\{a_{20}\}$ are shown, and
Figures 2A and 2B show the first sharp peak in $Re\{b_{31}\}$ as well as
$Im\{b_{31}\}$. These sharp peaks we call resonances. We notice that with
increasing n the first peak in each partial wave becomes narrower;
also with an increasing value of refractive index m the peaks become
narrower. At higher values of n not only the first peak but also a
few of the following peaks may be sharp enough to be called resonances.

2. WHAT ARE SURFACE WAVES?

There seems to be no generally accepted definition of what is a
surface wave in the case of scattering of electromagnetic waves on
small particles. We are going to define what we mean by a surface
wave. We have no intention in claiming that our definition should be
generally accepted. We simply feel that since we deal with surface
waves, we have to define clearly what is meant by it in this paper.

We make a reference to the two well-known phenomena generally
attributed to surface waves, namely the ripple structure of the extinction cross section (Fig. 3A) and to the glory phenomena of backscattering (Fig. 3B). From a large number of published papers and
from available monographs on light scattering we can learn that the
ripple structure in the extinction cross section is presumably caused
by the interference phenomena between a forward diffracted wave and
the surface wave. Similarly the peaks appearing in the backscattering
cross section (glory) are supposed to be a result of interference
between the backward reflected wave and the surface wave. Without
trying to discuss the validity of the above statements, we will show
that both the ripple structure of extinction cross section and the
glory in the backscattering are caused by the previously described
resonances in the partial wave amplitudes a_n and b_n.

Let us go back to Figures 2A, 3A and 3B. We notice that the
resonance in b_{31} (Fig. 2A) occurs somewhere between $26.9 < x < 27.0$.
We also notice that in the same region of x there appears a sharp
peak in the ripple structure (Fig. 3A) and a glory (sharp peak) in
the backscattering cross section (Fig. 3B). This leads to a conjecture, suggested some time ago by one of the authors, that there
is one-to-one correspondence between the partial wave resonances and
the ripple structure in the extinction and the glory in the backscattering.

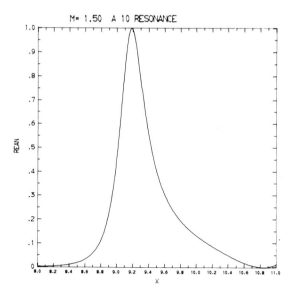

FIG. 1A.   The first resonance in the partial wave a_{10} occurs at
x = 9.203 for refractive index m = 1.50.

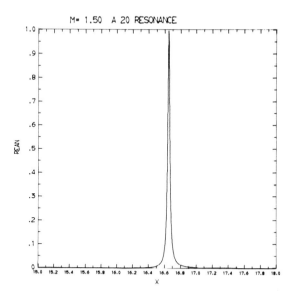

FIG. 1B.   With increasing n the resonances become narrower.   The
first resonance in a_{20} occurs at x = 16.65.

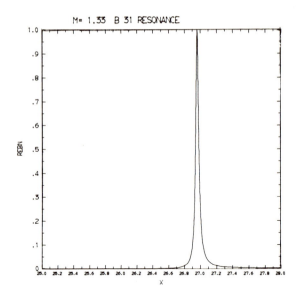

FIG. 2A.   In general the a_n resonances are sharper than the b_n resonances of the same value of n.   Also, with higher refractive index m, resonances are narrower.   The first resonance in b_{31} for refractive index m = 1.33 has a width comparable to the width of a_{20} (with m = 1.50).

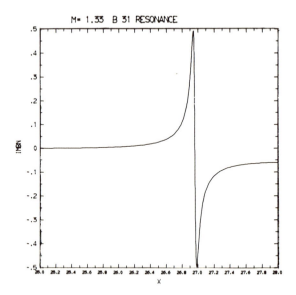

FIG. 2B.   The imaginary part of b_{31} goes through a zero in the direction from positive to negative values at the x value at which the real part of b_{31} reaches its maximum value $Re\{b_{31}\} = 1$.

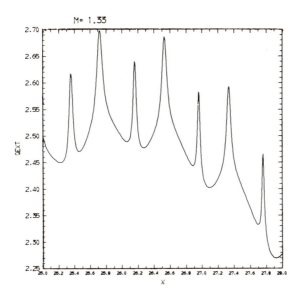

FIG. 3A.   The normalized extinction cross section Q_{EXT} (efficiency for extinction) shows a ripple structure.   Each peak in a ripple corresponds to a definite resonance in the partial waves a_n or b_n (surface waves).   The peak between $26.9 < x < 27.0$ corresponds to the b_{31} resonance.

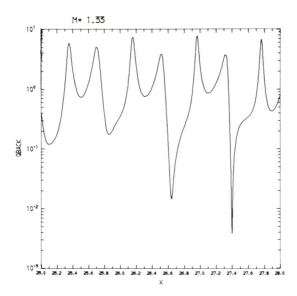

FIG. 3B.   The backscattering is completely dominated by the surface waves (partial wave resonances in a_n and b_n).   Again the peak in the backscattering between $26.9 < x < 27.0$ is caused by the first resonance in the b_{31} partial wave.

If our conjecture is correct, when we remove the resonance from the Mie calculations (set $b_{31} = 0$ in the resonance region $26.6 < x < 28.0$) we expect the peak in extinction and the glory in the back-scattering at $26.9 < x < 27.0$ to disappear. The results are shown in Figures 4A and 4B. Comparing Figure 3A with 4A and Figure 3B with 4B shows clearly that the peak in the ripple structure of the extinction wave between $26.9 < x < 27.0$ and the glory in the back-scattering in the same x region is caused by nothing else than the resonance in the partial wave amplitude b_{31} (Fig. 2A).

Since--as we have mentioned earlier--both the ripple and the glory are supposed to be caused by surface waves, and since we have shown that they are caused by nothing else than the resonances in the partial waves a_n and b_n, we will say that the resonances in the partial waves a_n and b_n represent a mathematical description of surface waves.

At higher values of x, it is possible that more than one partial wave resonates at the same value of x. Then we may say that the surface wave is a kind of collective phenomena arising from several partial waves resonating at the same size parameter x, as we have suggested earlier (Chýlek, 1976).

At the same time, we have to realize that there is a rather subjective judgment to decide when the width of a peak is narrow enough for a peak to be called a resonance.

3.   RESONANCE AND BACKGROUND CONTRIBUTION TO THE EXTINCTION CURVE

The value of the extinction at the peak of the ripple structure can be decomposed into two parts; the resonance contribution and the background contribution:

$$Q_{EXT}(x) = Q_{EXT,RES}(x) + Q_{EXT,B}(x) \qquad \text{where} \qquad (2)$$

$$Q_{EXT,RES}(x) = \frac{2}{x^2}(2N + 1)\,\text{Re}\{a_N(x)\} = \frac{2}{x^2}(2N + 1), \qquad (3)$$

$$Q_{EXT,B}(x) = \frac{2}{x^2}\sum_{n=1}^{\infty}{}'\,(2n + 1)\,\text{Re}\{(a_n + b_n)\}\ . \qquad (4)$$

We assume that it is the partial wave $a_n(x)$ that resonates at the considered value of x, and the symbol $\sum'$ means the sum over all contributing a_n and b_n amplitudes with the exception of the resonating wave a_N.

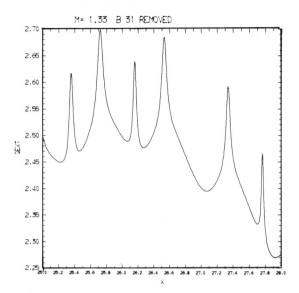

FIG. 4A.   To demonstrate that the ripple structure of the extinction curve is caused by the partial wave resonances in a_n and b_n, we have calculated again Q_{EXT}, however, now we have set $b_{31} = 0$ in the resonance region $26.6 < x < 27.4$. Comparing Figs. 3A and 4A we notice that the peak in Q_{EXT} between $26.9 < x < 27.0$ disappears, which shows clearly that the peak is formed by the b_{31} resonance.

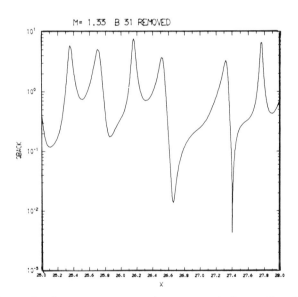

FIG. 4B.   When the b_{31} resonance is removed from the backscattering calculation the peak in the backscattering (glory) between $26.9 < x < 27.0$ disappears, which shows that the peak is formed by the b_{31} resonance.

Let us consider the resonance contributions of $a_{10}(x)$ and $a_{20}(x)$. The peak of the extinction curve at $x = 9.203$ (Fig. 5A) corresponds to the a_{10} resonance and the peak at $x = 16.65$ (Fig. 6A) corresponds to the a_{20} resonance. Using Equation (4) we can calculate the background contributions (see Fig. 5B and 6B) and we obtain $Q_{EXT,B}$ (9.203) = 1.982, $Q_{EXT,B}(16.65) = 2.502$. Subtracting the values from the extinctions, $Q_{EXT}(9.203) = 2.475$ and $Q_{EXT}(16.65) = 2.798$, we obtain for the resonance contributions $Q_{EXT,RES}(9.203) = 0.493$ and $Q_{EXT,RES}(16.65) = 0.296$ or, in the form of a ratio of the a_{20} and a_{10} resonance contributions, we obtain

$$\frac{Q_{EXT,RES}(16.65)}{Q_{EXT,RES}(9.203)} = 0.600 \ .$$

If we calculate the ratio of the same resonance contributions using an approximate form of equation (3) we obtain

$$\frac{Q_{EXT,RES}(16.65)}{Q_{EXT,RES}(9.203)} = \frac{41}{21} \frac{(9.203)^2}{(16.65)^2} = 0.596 \ ,$$

in excellent agreement with the previously obtained value of 0.600.

We conclude that the resonance or surface wave contribution to the extinction as described by equation (3) is in agreement with numerical results.

4.   SURFACE WAVES IN EXPERIMENTAL MEASUREMENTS

Recently a technique of optical levitation was used to support a liquid droplet in a laser beam (Ashkin, 1970; Ashkin and Dziedzic, 1976, 1977). One of the authors (A. Ashkin) used this technique in connection with a tunable laser to measure the backscattering. An example of the measurement of the backscattered intensity is shown in Figure 7A. Three strong peaks observed in the wavelength interval correspond to the glory effect. Since the droplet radius is known only approximately it is not possible to label the horizontal axis in size parameter units at this time.

The first point we want to demonstrate is that there is only one way to identify the experimental measurement with theoretical calculations even when the droplet radius is essentially unknown. The results of the backscattering calculations with the size parameter increment of $\Delta x = 0.001$ are shown in Figure 7B. We notice how the sharp resonance (glory) is superimposed on a broader background (really on the true background coming from the terms with $n < x$ plus

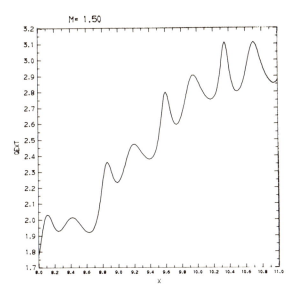

FIG. 5A.   The normalized extinction cross section Q_{EXT} is composed
from the resonance contributions superimposed on the background term.

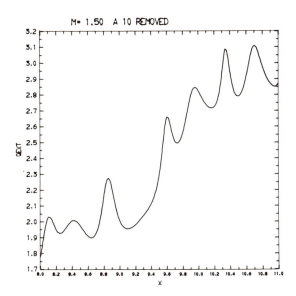

FIG. 5B.   When the first a_{10} resonance is removed around x = 9.2
only the background term remains.

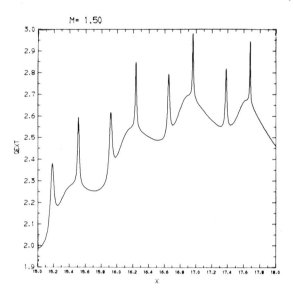

FIG. 6A.   The peak in Q_{EXT} around x = 16.6 is caused by the a_{20} resonance.

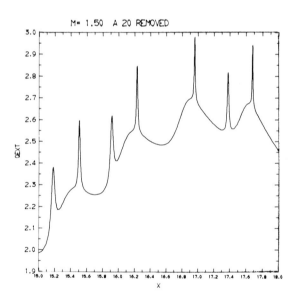

FIG. 6B.   When the a_{20} resonance is removed from the calculations (we set $a_{20} = 0$) the peak around x = 16.6 disappears and only the background term is left.

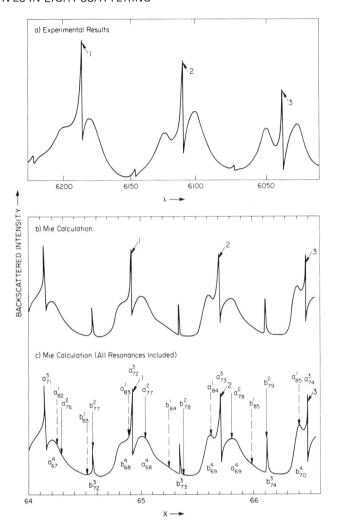

FIGS. 7A-7C. The experimental measurement (A) of the backscattering shows resonances superimposed upon the background. We notice narrow resonance peaks as well as the higher order broader resonances. The distance between the neighboring resonances of the same order depends slightly on the order of resonances and on the size parameter. Because of that, each group of resonances (denoted by 1, 2 and 3) has a different characteristic structure, $i.e.$, different morphology. The same structure appears in a coarse numerical calculation (B), and it is just this characteristic morphological structure of each group of resonances which allows us unambiguously to identify the experimentally observed peaks. A detailed numerical calculation (C) reveals additional narrow resonances that have not been observed experimentally because their width is below the resolution power of the experimental equipment used. The refractive index of oil droplet used is m = 1.42.

a broader higher order resonances in partial waves with n ≃ x). We conclude that the morphology of the resonance structure leads to an unambigous identification of the experimentally measured peak 1 in Figure 7A with the theoretically calculated peak 1 in Figure 7B. Now, since we know precisely the wavelength λ at which peak 1 was observed as well as the size parameter x at which peak 1 occurs in the numerical calculations, we can determine with a high degree of accuracy the radius of the observed droplet, as has been already demonstrated elsewhere (Chýlek *et al.*, 1978a, 1978b).

We were fully aware that by using the step size of $\Delta x = 0.001$ in our numerical calculations we could lose many of the narrow peaks in the backscattering structure. By using a computer resolution of the same order as the resolution of the experimental equipment used, we hoped to get comparable structures as shown in Figures 7A and 7B. By increasing the resolution of our numerical calculations we obtain the full backscattering structure (Fig. 7C). It follows that the resolution power of the instrumental arrangement used was not high enough to observe the first and the second order resonances. To observe these narrow resonances the resolution power of experimental observation has to be increased by several orders of magnitude.

5. SURFACE WAVES ON NON-SPHERICAL PARTICLES

Our preliminary results indicate that surface waves do exist at least in definite orientations of axially symmetric non-spherical particles. On the other hand, the strength of surface waves is considerably reduced when non-spherical particles are in random orientation. These conclusions were deduced from numerical calculations using the extended boundary condition method. Details of these calculations will be reported elsewhere.

Acknowledgements. The research was partially supported by the U.S. Army Research Office and the Atmospheric Research Section of the National Science Foundation. The National Center for Atmospheric Research is sponsored by the National Science Foundation.

REFERENCES

Ashkin, A., 1970, *Phys. Rev. Letters*, _24_, 156.
Ashkin, A. and Dziedzic, J. M., 1976, *Phys. Rev. Letters*, _36_, 267.
Ashkin, A. and Dziedzic, J. M., 1977, *Phys. Rev. Letters*, _38_, 1351.
Chýlek, P., 1973, *J. Opt. Soc. Am.*, _63_, 699.
Chýlek, P., 1976, *J. Opt. Soc. Am.*, _66_, 285.
Chýlek, P., Kiehl, J. and Ko, M., 1978a, *Phys. Rev.*, _A18_, 2229.
Chýlek, P., Kiehl, J. and Ko, M., 1978b, *Appl. Opt.*, _17_, 3019.

SHIFRIN'S METHOD APPLIED TO SCATTERING

BY TENUOUS NON-SPHERICAL PARTICLES

Charles Acquista

Drexel University, Dept. of Physics & Atmospheric Science

Philadelphia, Pa.

ABSTRACT

Shifrin constructed an integro-differential equation to describe
the scattering characteristics of an irregular particle and developed
an iterative procedure for solving the equation. This procedure is
fast enough to handle many physical problems that involve scattering
by a cloud of particles with different sizes, shapes and orientations.
In particular, the case of tenuous particles (small size or weakly
refracting) will be considered, and Shifrin's procedure will be shown
to be a generalization of the Rayleigh-Gans approximation.

A major goal of theoretical work on light scattering by irregular
particles is to provide a method of calculating the scattering matrix
for a collection of particles with different sizes and shapes and
random orientations. One method of attacking this problem is to
obtain an approximate scattering matrix for a single non-spherical
particle in a particular orientation and then to average this solution
over different orientations and different particle size parameters.
In general, this method provides greater accuracy for the collection
of particles as a whole than it does for any individual particle,
since the averaging described above tends to smooth out any sharp
individual structure. One drawback is that the averaging procedure
is so time-consuming that the single particle approximation must be
fast (and, therefore, simple). In order to keep the single particle
approximation simple, we will replace the collection of irregular
particles with a model collection of regular - but not necessarily
spherical - particles (e.g. cylinders, cubes or ellipsoids) which has

the same projected area and/or volume as the original collection.

Most of the aerosols and particulates in the atmosphere are tenuous in that they interact weakly with incident sunlight (or lidar illumination). Their refractive index is near 1 (typically 1.4), and they range in size up to a few microns. So, it is reasonable to build up a single particle scattering approximation in the form of a perturbation series about the incident field. Shifrin constructed an integro-differential equation which can be solved by the method of successive iterations. If the incident field is used as the first guess, Shifrin's method (Shifrin, 1968; Acquista, 1976) provides a type of Born series for the scattering matrix, the first iterate being identical to the Rayleigh-Gans approximation. A big advantage of this approximation is its simplicity. For most regular shapes, each term in the scattering matrix can be evaluated by means of a simple analytical formula which contains no integrations. However, the disadvantage of this approximation is that it is accurate only for small particles (e.g. if m = 1.5, the typical scatterer dimension should not exceed 0.01 μm). This disadvantage can be overcome by going to Shifrin's second iterate. Now the range of validity is increased tenfold but the approximation is considerably more compli-cated. Typically it involves a three dimensional integration with a removable singularity which must be evaluated numerically.

Using the second iterate, reasonable accuracy is achieved for scatterers with a typical dimension less than 0.1 μm for m = 1.5. For spherical (or nearly spherical) scatterers, this typical dimen-sion is the radius. For more irregular particles, the dimension of interest is the average path length through the particle. Thus, for a thin disk whose diameter is ten times its height, the dimension which governs the applicability of the formula is approximately $\sqrt{2}$ h. Even if such a disk had a diameter of 1 μm, the second iterate could be used successfully to calculate the scattering pattern.

Shifrin's itegro-differential equation for the effective electric field when an arbitrarily shaped particle is illuminated by the plane wave $E_o \exp(i\,k\cdot r)$ is

$$E_{eff}(r) = E_o e^{i\,k\cdot r} + \alpha(k^2 + \text{grad div}) \int U(r')\, E_{eff}(r') \times$$

$$\times\, G(r,r')\, d^3r' + \frac{4}{3}\pi\alpha U(r)\, E_{eff}(r) \quad,$$

where α is the polarizability,

$$\alpha = \frac{3(m^2-1)}{4\pi(m^2+2)} \ ,$$

G is the Green's function of the Helmholtz equation,

$$G(\mathbf{x},\mathbf{y}) = \frac{\exp(ik|\mathbf{x}-\mathbf{y}|)}{4\pi \ |\mathbf{x}-\mathbf{y}|} \ ,$$

and U is a pupil function or scattering potential defined by

$$U(\mathbf{r}) = \begin{cases} 1 \text{ if } \mathbf{r} \text{ is inside the scatter,} \\ 0 \text{ otherwise.} \end{cases}$$

The integro-differential equation can be changed into a pure integral equation by applying a Fourier transform, using the convolution theorem, and then inverting the transform. The end result is

$$\mathbf{E}_{eff}(\mathbf{r}) = \mathbf{E}_o e^{i\mathbf{k}\cdot\mathbf{r}} + \frac{\alpha}{(2\pi)^6} \iint e^{i\mathbf{p}\cdot\mathbf{r}} \ g(\mathbf{p}) \ u(\mathbf{p}') \times$$

$$\times [k^2 \mathbf{e}_{eff} (\mathbf{p}-\mathbf{p}') - \mathbf{p}(\mathbf{p}\cdot\mathbf{e}_{eff}(\mathbf{p}-\mathbf{p}'))] \ d^3p \ d^3p' + \frac{4}{3} \pi\alpha U(\mathbf{r}) \ \mathbf{E}_{eff}(\mathbf{r}) \ ,$$

where g, u and $\mathbf{e}_{eff}$ are the Fourier transforms of G, U and $\mathbf{E}_{eff}$. Constructing a power series for $\mathbf{E}_{eff}$ in terms of the small parameter α, we solve this equation by the method of successive iterations. In the first iteration we obtain

$$\mathbf{E}_{scat}^{(1)}(\mathbf{r}) = \frac{\alpha k^2}{r} \ e^{i\mathbf{k}\cdot\mathbf{r}} \ \mathbf{E}_{o\perp} \ u(k\hat{\mathbf{r}}-\mathbf{k}) \ ,$$

for the scattered field (effective field less incident field) far from the scatterer. Here $k\hat{\mathbf{r}}$ is the wave vector of the scattered radiation, $k\hat{\mathbf{r}} - \mathbf{k}$ is the momentum transfer, and $\mathbf{E}_{o\perp}$ is the component of $\mathbf{E}_o$ perpendicular to $\mathbf{r}$. When u is computed for spheres, disks or ellipsoids we find that $\mathbf{E}_{scat}^{(1)}$ is identical to the Rayleigh-Gans approximation.

In the second iteration we find

$$\mathbf{E}_{scat}^{(2)}(\mathbf{r}) = \mathbf{E}_{scat}^{(1)}(\mathbf{r}) + \frac{\alpha^2 k^2}{2\pi^2} \frac{e^{ikr}}{r} \int \frac{u(\mathbf{q}+k\hat{\mathbf{r}}) \ u(\mathbf{q}+\mathbf{k})}{q^2-k^2} \times$$

$$\times \left\{ (\frac{2}{3} k^2 + \frac{1}{3} q^2)\mathbf{E}_{o\perp} - (\mathbf{q}\cdot\mathbf{E}_o)\mathbf{q}_\perp \right\} \ d^3q \ .$$

These approximations for the scattered fields can be evaluated for arbitrarily shaped and oriented particles. In fact u, the transform of the pupil function or scattering potential, is the sole carrier of information about the scatterer. This function is readily obtainable for regularly shaped particles like disks, cylinders, spheroids, rectangular solids, etc. Once the scattered fields are known, the scattering matrix which relates the Stokes parameters of the incident and scattered radiation can be obtained in the usual manner (van de Hulst, 1957). For example, using the second iteration we obtain

$$
S_{11} = k^2 r^2 \frac{I_{scat}}{I_o}
$$

$$
= \alpha^2 k^6 u^2 (k\hat{r} - k) \left(\frac{1+\cos^2\beta}{2} \right) + \frac{\alpha^3 k^6}{\pi^2} u(k\hat{r} - k) \times
$$

$$
\times \int \frac{u(q+k\hat{r})\ u(q+k)}{q^2 - k^2} \left\{ (2k^2+q^2) \left(\frac{1+\cos^2\beta}{6} \right) - \right.
$$

$$
\left. - \frac{q^2}{2} \sin^2\theta\ (1-\cos^2\phi\sin^2\beta) + \frac{q^2}{8} \sin2\theta\ \sin2\beta\ \cos\phi \right\} d^3q ,
$$

where β is the scattering angle and θ, ϕ are the polar and azimuthal angles associated with the vector q.

We can use this approximation to model the polar nephelometer experiment of Holland and Gagne (1970) where a beam of partially polarized light illuminates a gas jet carrying particles of a known size and composition. The Stokes parameters of the scattered light are measured at different scattering angles to determine the scattering matrix S_{ij}. In view of electron microscope pictures of the scatterers, it seems reasonable to model them with a collection of oblate spheroids, all with the same minor axis and with the same projected area as the actual collection of particles. The predictions of this model are much closer to the experimental values of the scattering matrix than are calculations for a collection of equivalent spherical particles based on Mie theory.

REFERENCES

Acquista, C., 1976, *Appl. Opt.*, *11*, 2932.
Holland, A. C., and Gagne, G., 1970, *Appl. Opt.*, *9*, 1113.
Shifrin, K. S., 1951, "Scattering of Light in a Turbid Medium,"
  Moscow, (NASA Technical Translation TT F-477, 1968)
van de Hulst, H. C., 1957, "Light Scattering by Small Particles,"
  chap. 5, John Wiley and Sons, New York.

ENERGY CONSERVATION:

A TEST FOR SCATTERING APPROXIMATIONS

Charles Acquista

Drexel University, Dept. of Physics & Atmospheric Science

Philadelphia, Pa. 19101

A. C. Holland

NASA Wallops Flight Center

Wallops Island, Va. 23337

ABSTRACT

We explore the roles of the extinction theorem and energy conservation in obtaining the scattering and absorption cross sections for
several light scattering approximations. We show that the Rayleigh,
Rayleigh-Gans, Anomalous Diffraction, Geometrical Optics, and Shifrin
approximations all lead to reasonable values of the cross sections,
while the Modified Mie approximation does not. Further examination
of the Modified Mie approximation for the ensembles of non-spherical
particles reveals additional problems with that method.

Due to the complexity of solutions to the wave equation for non-
spherical scatterers, it is essential to make use of approximations
if we have any hope of determining the scattering pattern. Some
approximations are based on sound physical arguments. The Rayleigh
approximation, for example, is applicable when the scatterer dimensions are small compared to the wavelength of the incident light.
However, other approximations are based on more speculative arguments.
One way to check on the meaningfulness of a particular approximation
is to examine whether or not energy is conserved during the scattering
process. It sometimes happens that, after a simplifying assumption
is made, energy is no longer strictly conserved. This lack of energy

conservation often manifests itself in the form of unphysical results
for the scattering, absorption and extinction cross sections which
must always be non-negative and related by $C_{ext} = C_{scat} + C_{abs}$. We
will examine several scattering approximations to determine the method
of calculating the cross sections, the role of the extinction theorem,
and the extent to which energy is conserved in each approximation.
In particular, we will consider the Rayleigh, Rayleigh-Gans, Anomalous
Diffraction, Geometrical Optics, Shifrin, and Modified Mie approximations. We suggest that small excursions from strict energy conservation are understandable and insignificant; however, gross violations
of energy conservation cast doubt on the validity of the approximation
under consideration.

In a single scattering process, energy conservation requires that
the energy lost by an incident monochromatic light beam be equal to
the sum of the scattered and absorbed energies. These three energies
are obtained in different ways. Following van de Hulst (1957), we
express the distant scattered wave field by

$$u_s = S(\theta,\phi) \frac{\exp(ikr-ikz)}{ikr} u_o \quad ,$$

where u_o is an incident plane wave along the Z-axis and $S(\theta,\phi)$ is the
complex amplitude function. The total energy scattered into all
directions is then expressed as the energy of the incident wave
falling on a cross sectional area C_{scat} given by

$$C_{scat} = \frac{1}{k^2} \int_0^{2\pi}\int_0^{\pi} |S(\theta,\phi)|^2 \sin\theta \ d\theta \ d\phi \quad .$$

The energy lost by (or extinguished from) the incident beam can be
found by examining the interference between u_o and u_s in the forward
direction, $\theta = 0$. By evaluating $u_o + u_s$ it has been shown (van de
Hulst, 1957) that, far from the scatterer, the energy extinguished
from the incident wave is equal to the energy of the incident wave
falling on a cross sectional area C_{ext} where

$$C_{ext} = \frac{4\pi}{k^2} \text{Re}\big(S(0)\big) \quad .$$

This relation, which is called the extinction theorem in optics (or
the optical theorem in quantum mechanics), relates the energy removed
by absorption and scattering into *all* directions to the amplitude in
the *forward* direction. Energy conservation is not needed to establish
this theorem. At this point we can appeal to energy conservation to
calculate the absorption cross section from $C_{abs} = C_{ext} - C_{scat}$.
Classically, the amount of energy absorbed by the particle is found
by evaluating the net energy flux in a large sphere centered about
the particle. This calculation once again leads to the same result
for C_{abs} as we shall see. The net flux Φ is found by integrating the

imaginary part of $u^*\nabla u/k$ over the surface of a sphere:

$$\Phi = \frac{r^2}{k} \int_0^{2\pi}\int_0^{\pi} \mathrm{Im}(u_t\nabla u_t)\cdot\hat{e}_r \sin\theta \; d\theta \; d\phi \quad ,$$

where $u_t = u_o + u_s$. We can separate Φ into three terms. The first involves $u_o^*\nabla u_o$; it gives no contribution to the net flux. The second involves $u_s^*\nabla u_s$; it gives a contribution exactly equal to C_{scat}. The third term involves $u_o^*\nabla u_s + u_s^*\nabla u_o$; it gives a contribution

$$\Phi^{(3)} = \frac{r}{k} \int_0^{2\pi}\int_0^{\pi} \mathrm{Im}\left(S^*\exp(ikr - ikz)\right)(1 + \cos\theta)\sin\theta \; d\theta \; d\phi \quad ,$$

which is evaluated asymptotically in the limit $r \to \infty$. The rapidly varying exponential factor eliminates all contributions to this integral except near $\theta = 0$ where $r \simeq z + (x^2 + y^2)/(2z)$. This leads to a pair of Fresnel integrals which, when evaluated, yield $\Phi^{(3)} = - C_{ext}$. So, once again we have $\Phi = - C_{abs} = C_{scat} - C_{ext}$. In general, there is no way of calculating C_{abs} independently of C_{ext} and C_{scat}. Energy conservation is not an additional constraint; rather, it is a natural consequence of the theory. This analysis has been performed for scalar waves. The generalization to vector waves is immediate and straightforward (Born and Wolf, 1959; Kerker, 1969). In this case, the flux is replaced by the Poynting vector $\mathbf{P} = c \; \mathrm{Re}(\mathbf{E} \times \mathbf{H^*})/8\pi$. Recently, a controversy has arisen over the application of the extinction theorem to certain approximations to $S(\theta,\phi)$ which do not conserve energy (Acquista, 1978; Chýlek and Pinnick, 1979). It has been suggested that, in such cases, one should calculate the absorption by integrating over a sphere rather than by applying the extinction theorem. However, we have established here that both procedures are identical. We now use these formulae to calculate the cross sections in various approximations. We will evaluate the cross sections for a *spherical* scatterer of radius a, although they could be evaluated for any shape.

For the simplest form of the Rayleigh approximation (van de Hulst, 1957), the forward scattering function is

$$S(0) = ix^3\left[\frac{m^2 - 1}{m^2 + 2}\right] \quad ,$$

where $x = ka$ is the size parameter, and the scattering cross section is

$$C_{scat} = \frac{8\pi}{3k^2} x^6 \left|\frac{m^2 - 1}{m^2 + 2}\right|^2 \quad .$$

If we blindly follow the extinction theorem, we obtain $C_{ext} = 0$ for any real refractive index m. It appears as if the extinction theorem does not apply to Rayleigh scattering and, furthermore, it seems as

if we cannot calculate the absorption. But the Rayleigh approximation
is an asymptotic result with x as the small parameter. Thus, mathematically, we must evaluate C_{scat} and C_{ext} to the same order in x.
In fact, van de Hulst (1957) pointed out that $S(0)$ contains a real
term (when m is real) proportional to x^6 which is due to the effects
of radiation reaction. Using this more precise form of the Rayleigh
approximation, we find $Q_{ext} = Q_{scat}$ for real m. Thus the proper way
to apply the extinction theorem to Rayleigh's approximation in simple
form is to realize that the extinction theorem gives only absorption
in this case. So,

$$C_{abs} = \frac{4\pi}{k^2} \, \text{Re}\left(S(0)\right) = 4\pi ak^3 \text{Im}\left[\frac{m^2 - 1}{m^2 + 2}\right] \; .$$

Now, we obtain C_{ext} by adding C_{abs} to C_{scat}. The Rayleigh-Gans (van
de Hulst, 1957) and Shifrin (Shifrin, 1951; Acquista, 1976) approximations are treated in exactly the same way. For the Rayleigh-Gans
case, the absorption is the same as for the Rayleigh case but the
scattering is given by

$$C_{scat} = \pi a^2 |m-1|^2 \left\{ \frac{5}{2} + 2x^2 - \frac{\sin(4x)}{4x} - \frac{7}{16x^2}(1-\cos(4x)) \right.$$

$$\left. + \left(\frac{1}{2x^2} - 2\right)[\gamma + \ln(4x) - \text{Ci}(4x)] \right\} \; .$$

In Shifrin's approximation, the forward scattering amplitude is

$$S(0) = ix^3 \left\{ \left[\frac{m^2-1}{m^2+2}\right] + \left[\frac{m^2-1}{m^2+2}\right]^2 \left[\frac{7}{2} - \frac{9}{16}\left(\frac{4}{x} - \frac{1}{x^3}\right) \text{Si}(4x) \right. \right.$$

$$\left. \left. - \frac{63}{128x^3} \sin(4x) - \frac{9}{32x^2} \cos(4x)\right] \right\} \; ,$$

which is used to determine the absorption cross section, as above.
Si(x) and Ci(x), in the equations above, refer to the Sine and Cosine
Integrals, respectively, and γ is Euler's constant). The scattering
cross section can only be obtained by integration (Acquista, 1976).
In the Geometrical Optics and Anomalous Diffraction approximations,
the incident light can be separated into rays. In these cases there
are no problems with unphysical results for the cross sections when m
is real. For the Geometrical Optics case, $S(0) = x^2/2$ and, by
applying the extinction theorem, $C_{ext} = 2\pi a^2$. By introducing Fresnel
coefficients and summing over the contribution of each ray we obtain
for $|m| > 2$,

$$C_{scat} = \pi a^2 \left[1 - \frac{f(p) + f(q)}{2}\right] \; ,$$

where

$$p = \sqrt{\left(-2/\mathrm{Im}(m^2)\right)} \quad ,$$

$$q = \sqrt{\left(-2\,\mathrm{Im}(m^2)\right)} \quad ,$$

and

$$f(x) = 8[x - \ln(1 + x + x^2/2)]/x^2 .$$

For smaller values of $|m|$ the Anomalous Diffraction approximation is used. In this case

$$S(0) = x^2\left\{\frac{1}{2} + \frac{1}{\delta^2} + \frac{(i\delta - 1)}{\delta^2}\,\exp(-i\delta)\right\}$$

where $\delta = 2ka(m - 1)$. Applying the extinction theorem, the extinction cross section is given by

$$C_{ext} = 4\pi a^2\left\{\frac{1}{2} + \mathrm{Re}[\,\frac{1}{\delta^2} + \frac{(i\delta - 1)}{\delta^2}\,\exp(-i\delta)]\right\} \quad .$$

In this case summing over all rays to find C_{scat} is difficult. Van de Hulst (1957) has given a short cut for calculating C_{abs} directly, based on the applicability of the ray approximation and the smallness of $|m - 1|$:

$$C_{abs} = \left\{\frac{1}{2} + \frac{e^{-w}}{w} + \frac{e^{-w}-1}{w^2}\right\}2\pi a^2$$

where $w = 2\,\mathrm{Im}(\delta)$.

The Modified Mie approximation (Chýlek, et al., 1976) is unlike previous approximations in that it does not apply to single scatterers (but rather to ensembles of randomly oriented and arbitrarily shaped non-spherical particles), and in that it does not rely on the large-ness or smallness of the parameters x, $|\delta|$, or $|m - 1|$. Instead it relies on the conjecture that the absence of surface waves is the main difference between scattering by irregular particles and scattering by spheres. The proposed modification to Mie theory consists in ad-justing the Mie coefficients in a way which has been shown to violate energy conservation (Acquista, 1978). Using the computations of Welch and Cox (1978), we can demonstrate that Modified Mie theory leads to serious differences between C_{ext} and C_{scat} when m is real. In fact, they reported that, for certain values of x, $C_{ext} \simeq 2C_{scat}$ - a gross violation of energy conservation. The situation here ($C_{ext} > C_{scat}$) is quite different from the first application of the extinction theorem in the Rayleigh approximation where $C_{ext} < C_{scat}$. Now we have too much absorption, and, aside from the mathematical reasons for this which have already been given (Acquista, 1978), we should explore

the physical reasons. First, this approximation is based on the conjecture that, in comparing the scattering characteristics of spherical
particles with irregular particles, only the surface waves are
different. Implicit in this conjecture is the essential equivalence
of the transmitted and reflected waves in both cases. However, in
view of the geometrical fact that a sphere is the shape which mini-
mizes the ratio of surface area to volume, one should expect significant differences in the scattered and absorbed intensities due to
differences in the transmitted and reflected waves unless quantitative calculations show otherwise. Second, in this approximation it
has not been proved that the surface waves have been eliminated. In
fact, the only connection between the peaks in the Mie coefficients
a_n and b_n which are modified and surface waves is that, for each
coefficient, the peak occurs near $x = n$ (Chýlek, 1976). Here, n is
the summation index in the Mie partial wave expansion. Furthermore,
it can be argued that, even after the modification, the ripple
structure in the extinction cross section (which has been attributed
to surface waves in general and the peaks in the coefficients in
particular) still does not vanish. If we examine the size of the
ripple structure published earlier (Chýlek, 1976) near $x = 10$ and
near $x = 20$ we see that there is no reduction in intensity. However,
if these structures were due only to individual peaks in the coefficients a_{10} and a_{20}, we would expect the second peak to be half the
size of the first since C_{ext} can be written as

$$C_{ext}(x) = \frac{2\pi a^2}{x^2} \sum_{n=1}^{\infty} (2n + 1)\operatorname{Re}(a_n + b_n) \quad ,$$

and at the peak, $x \simeq n$.

By calculating the cross sections for real values of the refractive index, we have shown that there are serious problems with
the Modified Mie approximation to $S(\theta,\phi)$, especially near $\theta = 0$.
There is, however, another objection to this approximation which
concerns its domain of validity. We are told that this approximation
applies to "arbitrarily shaped non-spherical particles," (Chýlek,
et al., 1976) although no definition is given as to what constitutes
an arbitrary shape. Does this approximation apply to an ensemble of
flat, disklike, irregular particles, or do some of the particles have
to be chunky? Does it apply to salt particles which tend to exhibit
a cubical structure, or to ice crystals, or to aerosols with a water
jacket? These and related issues should be addressed before results
based on this or any other approximation can be accepted.

Comment (WEIL): It isn't correct to say that there is no way to
avoid using the extinction (optical) theorem to find the absorption.
You can directly compute the Joule heating in the particle:

$$\int_{\substack{\text{Volume of} \\ \text{scatterer}}} \sigma E^2 dv$$

where σ = conductivity.

Reply (ACQUISTA): I agree that there are other methods for calculating the absorption cross section; however, all methods lead to
the same result for C_{abs} (whether this result is meaningful or not).
In particular, the method that you refer to involves integrating the
divergence of the local flux (the Poynting vector) over the volume of
the scatterer. This is mathematically equivalent to the surface
integral I used because, by the Gauss integral theorem,

$$\int \nabla \cdot \Phi dv = \int \Phi \cdot \hat{e}_r d\Omega \ .$$

(These integrals can be performed either over the particle or over a
large volume containing the particle). So, although there may be ways
of avoiding the *use* of the optical theorem, there is no way to avoid
the *consequences* of that theorem. In the modified Mie approximation,
the method you suggest for calculating the absorption cannot be
applied since only the far fields are calculated and the electric
field inside the scatterer remains unknown.

REFERENCES

Acquista, C., 1976, *Appl. Opt.* *15*, 2932.
Acquista, C., 1978, *Appl. Opt.* *17*, 3851.
Born, M., and Wolf, E., 1959, "Principles of Optics," Pergamon,
  Oxford, 657–659.
Chýlek, P., 1976, *J. Opt. Soc. Amer.* *66*, 285.
Chýlek, P., Grams, G. W., and Pinnick, R. G., 1976, *Science 193*, 480.
Chýlek, P., and Pinnick, R. G., 1979, *Appl. Opt.* *18*, 1123.
Kerker, M., 1969, "The Scattering of Light and Other Electromagnetic
  Radiation," Academic, N. Y., 49–50.
Shifrin, K. S., 1951, "Scattering of Light in a Turbid Medium,"
  Moscow, (NASA Technical Translation TT F-477, 1968).
van de Hulst, H. C., 1957, "Light Scattering by Small Particles,"
  Wiley, N.Y., see especially pages 12, 29–31, 66, 70, 88, 175, and
  289–292.
Welch, R. M., and Cox, S. K., 1978, *Appl. Opt.* *17*, 3159.

ELECTROMAGNETIC SCATTERING FROM TWO IDENTICAL PSEUDOSPHERES

George W. Kattawar and Terry J. Humphreys

Texas A&M University, Department of Physics

College Station, Texas   77843

ABSTRACT

In an attempt to gain some insight into the problem of how close particles have to be on the average before phase effects become important, we have calculated the cross section and phase matrices for two identical particles as a function of the distance between their centers for both broadside and end-on illumination. We employed the point dipole approximation to build two pseudospherical particles each consisting of 32 dipoles with an effective size parameter of 0.9283 and refractive index of $1.54-0.0i$.

In regard to cross sections we found that in certain cases several percent deviation from the converged value, which is twice the single particle cross section, could be found up to distances of separation of 25 single particle diameters. The maxima and minima in the cross section curves can be explained in terms of simple interference effects. Also the element P_{11} of the phase matrix showed a great deal of structure, and the position of maxima and minima could be explained by phase shift analysis.

1. THEORETICAL FOUNDATION

The method we have employed was first introduced by Purcell and Pennypacker (1973). The basic idea behind the method is to treat a continuum as a collection of point dipoles which interact both with the external field and each other. The advantage of such a scheme is the fact that we are no longer trying to solve a boundary value problem which is why we can build irregular objects as easily as regular objects.

In what is to follow we will employ the Gaussian system of units. Let us consider a right handed coordinate system with unit vectors 1_x, 1_y and 1_z such that $1_x \times 1_y = 1_z$ with cyclic permutation of x, y and z. We will define the scattering plane to be the x - z plane and will also adopt the notation that the subscripts $\parallel$ represents components in the x - z plane whereas $\perp$ will denote components perpendicular to the x - z plane. Let us now consider a plane electromagnetic wave of unit amplitude propagating along the positive x direction and polarized along the z axis, $i.e.$,

$$E_{inc} = 1_z \, e^{i(kx-\omega t)} \tag{1}$$

where $k = 2\pi/\lambda$ and $c = \omega/k$. Now a polarizable particle situated at position r_i will see, in addition to the applied external field, the field radiated from all the other dipoles in the medium. Therefore, the local field E_i seen by the ith dipole is

$$E_i = 1_z \, e^{ikx_i} + \Sigma_{j \neq i} \left\{ \frac{\exp(ikr_{ij})}{r_{ij}^3} \left[k^2 (r_{ij} \times p_j) \times r_{ij} + \frac{(1-ikr_{ij})}{r_{ij}^2} \right] \right. \tag{2}$$

$$\left[(3p_j \cdot r_{ij}) r_{ij} - r_{ij}^2 p_j) \right] \Bigg\} ,$$

where r_{ij} is the distance of separation between the ith and jth dipoles.

The individual dipole amplitudes p_i are related to the local field E_i by

$$p_i = \alpha_i E_i \tag{3}$$

where α_i is the polarizability of the ith dipole. Although we have assumed the polarizability to be isotropic, this formulation can easily be extended to handle anisotropic cases. The connection between the polarizability and the refractive index, n, is obtained through the Classius-Mossotti relation, namely

$$\alpha = \frac{3(n^2-1)}{4\pi N(n^2+2)} \tag{4}$$

where N is the number of dipoles per unit volume. It is important to note that this relation is exact for point dipoles on a cubic lattice. With the polarizability determined from equation (4), equations (2) and (3) can be set up in matrix form to solve for the $\mathbf{p_i}$, *i.e.*,

$$\underset{\sim}{B} \, \underset{\sim}{p} = \underset{\sim}{E}_{inc} \, . \tag{5}$$

The matrix $\underset{\sim}{B}$ is a $3M \times 3M$ complex, symmetric matrix where M is the total number of dipoles. To solve this equation we use the well known conjugate gradient method which is an iterative technique which guarantees convergence, within roundoff error, when the number of iterations equals the order of the matrix. Once the dipole amplitudes have been determined then we can compute the far field components by summing the far field results for all the radiating dipoles. We can now define the complex vector scattering amplitude $\mathbf{A}(\theta,\phi)$ by

$$\mathbf{E} = \mathbf{1}_z e^{ikx} - \frac{e^{ikr}}{ikr} \, \mathbf{A}(\theta,\phi) \, . \tag{6}$$

Since we now have all components of the vector field, we can easily transform it in directions parallel and perpendicular to the scattering plane. This allows us to use the notation of van de Hulst (1957) as follows:

$$\begin{bmatrix} E_{||} \\ E_{\perp} \end{bmatrix} = \begin{bmatrix} A_2 & A_3 \\ A_4 & A_1 \end{bmatrix} \begin{bmatrix} E_{inc} \\ E_{inc} \end{bmatrix} \, . \tag{7}$$

With this notation the phase matrix $\underset{\sim}{P}$ assumes the following form:

$$\underset{\sim}{P} = \begin{bmatrix} \dfrac{M_1+M_2+M_3+M_4}{2}, & \dfrac{M_2-M_3+M_4-M_1}{2}, & S_{23}+S_{41}, & -D_{23}-D_{41} \\[2mm] \dfrac{M_2+M_3-M_4-M_1}{2}, & \dfrac{M_2-M_3-M_4+M_1}{2}, & S_{23}-S_{41}, & -D_{23}+D_{41} \\[2mm] S_{24}+S_{31}, & S_{24}-S_{31}, & S_{21}+S_{34}, & -D_{21}+D_{34} \\[2mm] D_{24}+D_{31}, & D_{24}-D_{31}, & D_{21}+D_{34}, & S_{21}-S_{34} \end{bmatrix} \tag{8}$$

where $M_k = A_k A_k^*$

$$S_{kj} = S_{jk} = (A_j A_k^* + A_k A_j^*)/2 \tag{9}$$

$$-D_{kj} = D_{jk} = i(A_j A_k^* - A_k A_j^*)/2 \, ,$$

where k, j = 1, 2, 3, 4 and the asterisk denotes complex conjugation.

The extinction cross section for say the $||$ (z-component) can be obtained by

$$\sigma_{ext_{||}} = \frac{4\pi}{k^2} \ Im \ (\mathbf{1}_z \cdot \mathbf{A})_{\theta=0^\circ} \tag{10}$$

where $(\mathbf{1}_z \cdot \mathbf{A})_{\theta=0^\circ}$ denotes the z-component of the scattered far field in the forward direction. A similar result holds for the incident wave polarized along the y-axis.

The absorption cross section for either component can be obtained by

$$\sigma_{abs} = \frac{4\pi k \alpha''}{|\alpha|^2} \sum_{i=1}^{M} (|p_{xi}|^2 + |p_{yi}|^2 + |p_{zi}|^2) \tag{11}$$

where α'' is the imaginary part of the complex polarizability, *i.e.*, $\alpha = \alpha' + i\alpha''$. Our calculation scheme is now complete since we can compute all measurable parameters.

2.   TEST RESULTS

To test the accuracy of the method we built a sphere of cubic lattices with a total of 136 dipoles (see Figure 1) similar to the one used by Purcell (1973). All distances in the calculation were expressed in units of d which is the interdipole separation distance. We equated the volume of this pseudosphere, in this case $136d^3$, to the volume of a sphere of radius r_{eff} by

$$r_{eff} = (\frac{3 \times 136}{4\pi})^{1/3}d = 3.1902 \ d$$

We chose the wavelength to be 20.045d so as not to produce too large a phase shift between adjacent dipoles and at the same time produce an effective size $x_{eff} = k \ r_{eff} = 1.0$. We chose a real refractive index n = 1.33 . For comparison we also performed the Mie calculation for a sphere of the above effective size and index of refraction. It is important to note that with this formulation there are *no* adjustable parameters. In Figures 1 and 2 we show a comparison between the Mie theory and the dipole approximation for two elements of the phase matrix, namely P_{11}, and P_{12}/P_{11}, respectively. As can be seen the agreement is extremely good. The worst errors are only of the order of 1%. The extinction coefficient, $Q_{ext} = \sigma_{ext}/(\pi \ r_{eff}^2)$, also agrees to about 1.3%. Similar accuracy was also achieved on

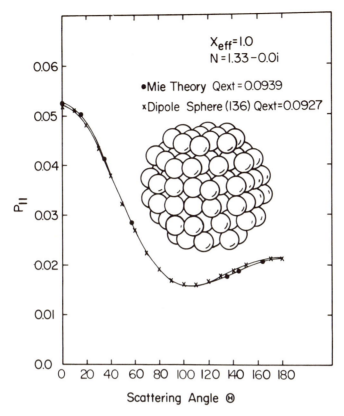

FIG. 1.   Comparison of the phase matrix element P_{11} as a function of
scattering angle for a sphere of effective size x_{eff} = 1.0 and refractive index 1.33-0.0i with a 136 dipole sphere (shown in the
inset).   The extinction coefficient $Q_{ext} = \sigma_{ext}/(\pi r^2_{eff})$ is also
shown for each.

elements P_{33}/P_{11} and P_{43}/P_{11}.   The results of this calculation have
given us a great deal of confidence in this formulation.

3.   CALCULATIONAL RESULTS

To study the behavior of the cross sections and phase matrix on
the separation of particles we constructed the following model.   We
constructed two symmetric particles, hereafter referred to as pseudospheres, each consisting of 32 dipoles (see Figure 3).   The refractive index was chosen to be 1.54-0.0i corresponding to such things
as NaCl at 6328Å and $(NH_4)_2SO_4$ at 3250Å.   We can also define a sphere
for each particle having the same equivalent volume by

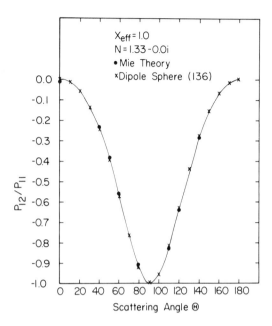

FIG. 2.   Same as Fig. 1 except we compare the reduced phase matrix element P_{12}/P_{11}.

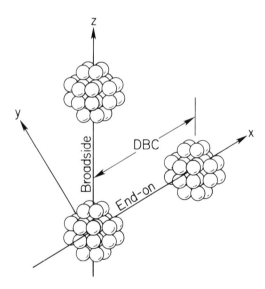

FIG. 3.   Structure and configuration for the two 32 dipole pseudo-spheres showing broadside and end-on incidence.

$$r_{eff} = \left(\frac{3 \times 32}{4\pi}\right)^{1/3} d = 1.969 \ d$$

where d is the interdipole separation. The wavelength was set at 13.3302d producing an effective size parameter $x_{eff} = 2\pi r_{eff}/\lambda = 0.928$ which puts us out of the Rayleigh regime. In Figures 4 and 5 we compare the elements of the phase matrix namely P_{11}, and P_{12}/P_{11}, respectively with the corresponding elements computed from Mie theory.

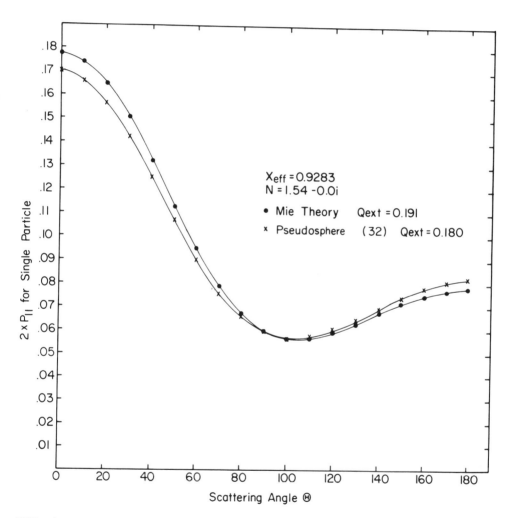

FIG. 4.   A comparison of the phase matrix element $2P_{11}$ as a function of scattering angle, for a sphere of effective size x_{eff} = 0.9283 and refractive index 1.54-0.0i, with a 32 dipole pseudosphere. Q_{ext} is also shown for each.

G. W. KATTAWAR AND T. J. HUMPHREYS

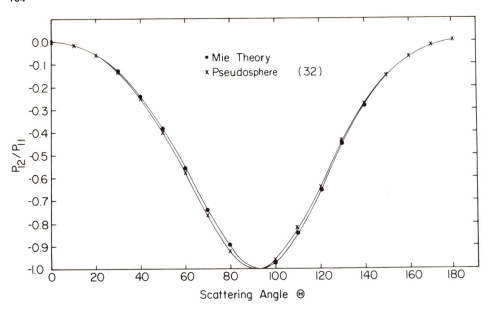

FIG. 5.   Same as Fig. 4 except we compare the reduced phase matrix element P_{12}/P_{11}.

The worst disagreement is of the order of 5%. This discrepancy is due to the lumpiness of our pseudosphere. However, there is no cause for concern since all comparison for two particle scattering versus single particle scattering will be for the pseudospheres.

The two configurations we considered are shown in Figure 3. One is referred to as broadside incidence, when the two particles are aligned along the z-axis, while the other is referred to as end-on incidence where the particles are aligned along the x-axis. The distance between centers (DBC) is also depicted in the diagram. When DBC = 4 (in units of inter-dipole separation) the pseudospheres are touching. In Figure 6 we show the extinction cross sections $\sigma_{\perp}$ (incident field polarized along the y-axis) and $\sigma_{||}$ (incident field polarized along the z-axis) for both broadside and end-on incidence as a function of the distance between centers (DBC) of the pseudospheres. For comparison we also show the value of twice the single pseudosphere cross section ($2\sigma_s$) which is what we would obtain if phase effects are neglected. For the case of broadside incidence, $\sigma_{||}$ is slightly more than four times the single pseudosphere cross section but falls rapidly as the particles move apart and converges rapidly to twice the single pseudosphere value. It is interesting to note that the distances between maxima and minima are very closely one wavelength (13.3302d) apart which is precisely what one would

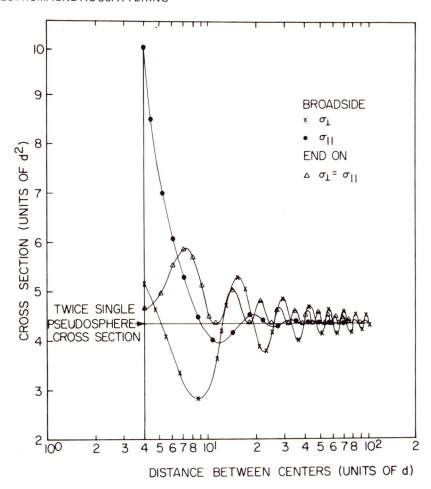

FIG. 6.   Cross section versus distance between centers (DBC) for two
pseudospheres for both broadside and end-on incidence. $\sigma_{||}$ and $\sigma_{\perp}$
are the cross section for the incident field polarized along the z
and y axis respectively.

expect from simple interference effects. Now this behavior should be
contrasted with the behavior of $\sigma_{\perp}$ which starts out at roughly half
the value of $\sigma_{||}$ but converges much less rapidly. In fact, even out
to DBC = 100 the maxima are still deviating by several percent over
the converged value. The reason for the difference in behavior can be
explained by the following simple model. When the incident wave is
polarized along the y-axis, the radiating dipoles are also effectively
aligned along the y-axis. Therefore, since the two particles are
aligned along the z-axis, which is perpendicular to the dipole axis,
they see quite intense fields from each other. When the field is

polarized along the z-axis, the induced dipoles are essentially
aligned along the z-axis and therefore the radiation that one sees
from the other is greatly diminished. Hence, their mutual interaction
is reduced. For the case of end-on incidence $\sigma_\perp = \sigma_{||}$. There are
two noteworthy features about this case. First the convergence is
slow and occurs from above compared to $2\sigma_s$. Secondly, the maxima and
minima are separated by approximately a half wavelength. The reason
for this is as follows. For broadside incidence the corresponding
dipoles of the two particles are stimulated in phase whereas for the
end-on incidence there is a phase lag between stimulation of the first
and second particles. Therefore, we should expect twice as many
maxima and minima compared to the case of broadside incidence and
this is precisely what we observe in Figure 6. In Figure 7 we present the element P_{11} of the phase matrix, often referred to as the

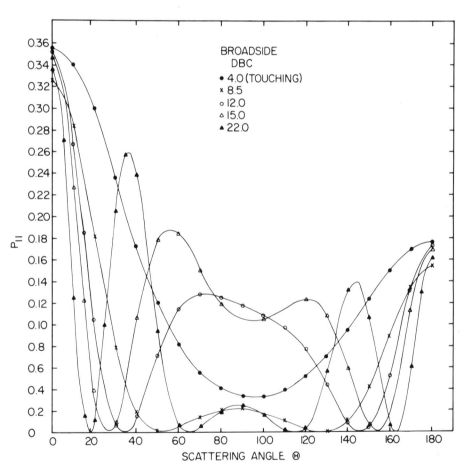

FIG. 7. P_{11} versus scattering angle for two pseudospheres at various
distances of separation for broadside incidence.

phase function, for broadside incidence for selected values of DBC. We first note that when DBC = 4.0, particles touching, that the ratio of the maximum to minimum value is approximately 11, whereas for the single pseudosphere (see Fig. 4) it is only 3. As the particles become separated, maxima and minima begin to develop. These can again be understood in terms of two sources radiating in phase. When the scattering angle Θ is such that the phase difference is an odd integral multiple of π we should expect destructive interference, *i.e.*, when

$$\sin \Theta = (2m + 1)\lambda/2D \qquad (12)$$

where m can be any integer, including zero, such that

$$m \leq \left(\frac{2D}{\lambda} - 1 \right)/2 \qquad (13)$$

and D is the DBC. Let us consider the case where D = 22.0d; then m = 0, 1. From equation (12) we expect *minima* to occur at scattering angles of 17.6, 65.3, 114.6 and 162.4°. From Figure 7 we see that these are precisely the angles where the minima occur. A similar analysis for the maxima (with phase shifts of integral multiples of 2π) predicts them to be at scattering angles of 33.7 and 46.3° which is again in agreement with the values shown in Figure 7. Another point worth noting is that backscattering ($\Theta = 180°$) is fairly insensitive to particle separation and is essentially four times the value for a single particle which is what one would expect when the dipoles are radiating in phase. In Figure 8 we show the corresponding case for end-on incidence. Due to the additional phase lag between stimulation and emission, equation (12) for the minima now becomes

$$\cos \theta = 1 - \frac{(2m+1)\lambda}{2D} \qquad (14)$$

where

$$m \leq \left(\frac{2D}{\lambda} - 1 \right)/2 \ . \qquad (15)$$

Again using D = 22.0d as an example and using equation (14) we predict minima at 45.8, 84.8, and 121°. From Figure 8 we see that these are precisely the minimum points. A similar analysis predicts the maxima to be at 66.8, 102.2, and 144.9 which is also in agreement with the curve in Figure 8. In Figure 9 we show the ratio of P_{12}/P_{11}

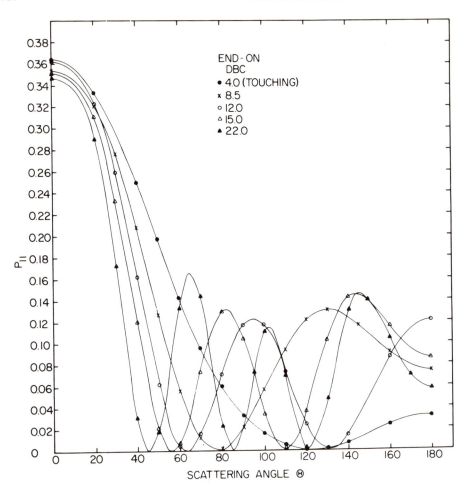

FIG. 8.   Same as Fig. 7 except for end-on incidence.

the quantity normally measured experimentally, for broadside incidence.   These curves can be compared to the single pseudosphere curve
shown in Figure 5.   We first note that if the pseudospheres are
touching there is a marked difference when compared to the single
pseudosphere case; however, as the particles are moved apart this
difference disappears.   Also the sharp spikes which appear occur at
the precise angles where P_{11} reaches its minima.   This case can be
compared with Figure 10 which is for end-on incidence.   Here we see
that the interference effects are much weaker when compared to the
case of broadside incidence.   The same qualitative effects can also
be seen in comparing the other elements of the phase matrix.

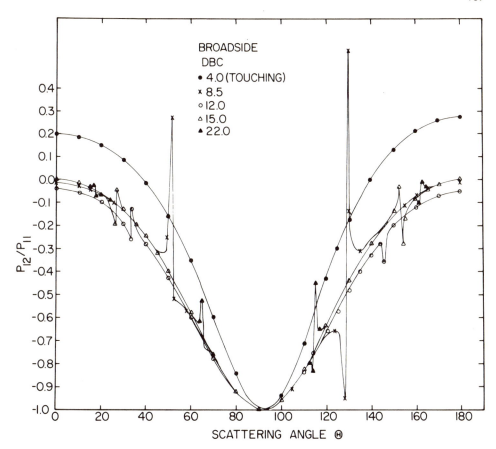

FIG. 9.   Same as Fig. 7 except for the reduced phase matrix element
P_{12}/P_{11}.

4.   CONCLUSION

We have seen that the point dipole approximation can be effectively used to study cross sections and phase matrices of irregularly
shaped objects.   The method can also be used to handle particles
which are birefringent.   Another extremely important aspect is that
incident fields other than plane waves can be used.   This is extremely important in laser optical levitation studies.

Acknowledgements.   We would like to thank our colleague, Dr.
Randall C. Thompson, for many helpful discussions.   This work was
supported by Grant No. NGR-44-001-117 from the National Aeronautics
and Space Administration.

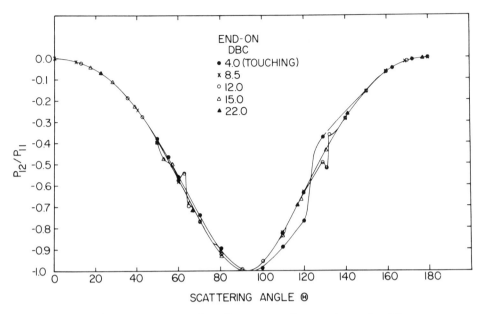

FIG. 10.   Same as Fig. 9 except for end-on incidence.

Question (KERKER):   Have you compared your results with the limiting solution of the "Trinks" problem such as published by Levine and Olaofe.

Answer (KATTAWAR):   We have compared our results with the results of Levine and Olaafe and the agreement was quite good. It should be emphasized, however, that they only considered two dipoles.

REFERENCES

Purcell, E. M., and Pennypacker, C. R., 1973, *Ap. J.*, <u>186</u>.
van de Hulst, H. C., 1957, "Light Scattering by Small Particles,"
  Wiley, N. Y.

ABSORPTION BY PARTICLE EDGES AND ASPERITIES

J. R. Aronson and A. G. Emslie

Arthur D. Little, Inc.

Cambridge, Massachusetts   02140

ABSTRACT

Measurements of infrared emittance from powders indicate that edges and other asperities on the particles cause enhanced absorption. A theory of the particle phase function, in which the asperities are represented by a surface distribution of induced dipoles, was able to fit the experimental results and to predict the outcome of further experiments correctly.

For a number of years we have been working on the development of a theory of the spectral reflectance or emittance of particulate materials composed of irregularly shaped particles (Emslie and Aronson, 1973; Aronson and Emslie, 1973, 1975). The theory has developed by the successive refinement of approximations as required by experimental measurements of the infrared spectra of powders.

During the course of this work we observed that for both corundum (Figure 1) and quartz the reflectance varies in opposite directions with particle size in spectral regions of high transparency and opacity. In particular, in opaque regions such as occur in Reststrahlen bands the reflectance is diminished as the particle size becomes finer. Another phenomenon is the frequent appearance of a minimum in the center region of strong Reststrahlen bands. Neither of these results could be explained by a theory based on smooth particles. These phenomena did not occur in the case of smooth spherical glass powders (Figure 2), which provided our first clue as to a mechanism.

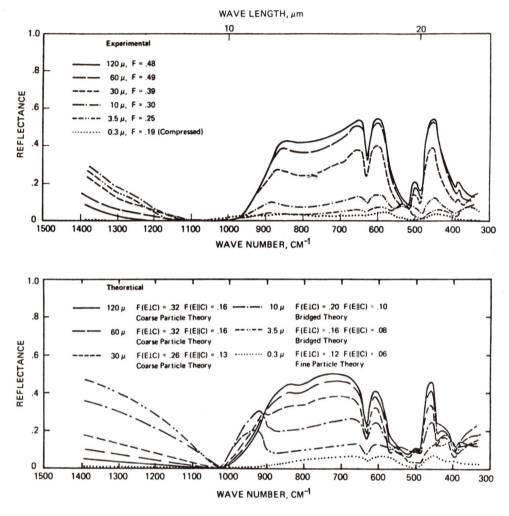

FIG. 1.   Theoretical predictions and experimental results for a wide range of particle sizes of corundum powders. F denotes the volume fraction; E⊥C means the electric vector is perpendicular to the C axis.

     Scanning electron micrographs (Figures 3a and 3b) of our parti-
cles showed significant numbers of asperities on the particle surfaces. Of course all jagged particles have regions, namely those of
edges and corners, where asperity effects might be expected to occur.
In fact, were there no *surface* asperities, one could predict *a priori*
that a larger proportion of a small particle consists of edge and
corner regions than of a large particle. This was precisely the

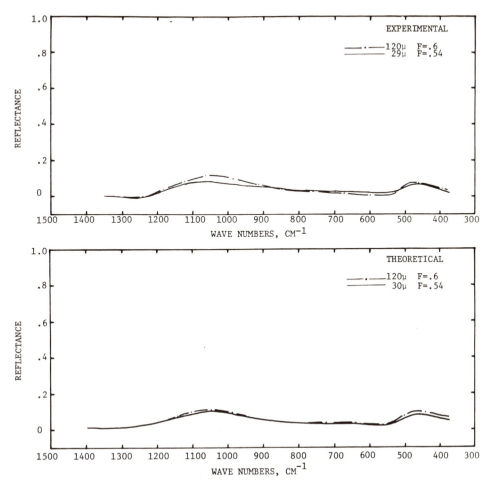

FIG. 2.   Comparison of experimental and theoretical reflectance of glass beads.

kind of effect needed to explain the particle size dependence of the Reststrahlen bands, and so the hypothesis was made that additional absorption occurs in the regions of sharp corners, edges, and surface asperities. An additional clue was found in that small amounts of very fine particles clinging to larger particles produced the same sort of experimental feature as the surface asperities did (Figure 4).

An addition to our smooth particle theory was then constructed

FIG. 3a.   Scanning electron micrograph of steplike asperities.   These
corundum particles have a characteristic size of 30 μ. 1000x

FIG. 3b.   Scanning electron micrograph of steplike asperities and
additional fines.   These corundum particles have a characteristic
size of 120 μ. 500x

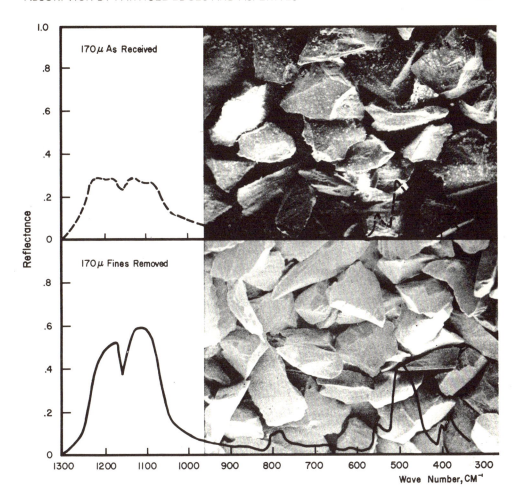

FIG. 4. Reproduction of cover of Applied Optics journal, dated
November 1973. The large reflectance differences shown in the infrared Reststrahlen regions when a small proportion of clinging fines
are removed from a particulate sample of quartz are explained by
ADL theory of reflectance discussed in the journal's feature articles.

assuming that the effects of the asperities and edges could be modeled
by a surface distribution of induced dipoles, with an adjustable parameter being necessary in the case of the surface asperities simply to
provide for the number and size of asperities present. In the case
of edges we assumed that the effective depth of the edge is related to
the penetration depth of the wave and the effective width is proportional to the free space wavelength (Emslie and Aronson, 1973). When
these phenomena were then modeled in our computer program, we found

that we could indeed fit the various measurements of powder samples.

In order to substantiate our hypothesis we conducted several additional experiments. We first took a single crystal of sapphire, measured its emittance (emittance = 1-reflectance) spectrum and roughened it with diamond paste. After washing the surface clean of any clinging material, we remeasured the emittance spectrum and found that as predicted the characteristic feature had been caused to appear by the roughening process although it is not present for the polished sample (Figure 5). A second experiment was carried out in which emittance measurements of spherical particle corundum powders were made and compared with those for the usual chunky or platelet habit corundum powders. For the same particle sizes it was observed that the emittance level was changed (compare Figure 6 with Figure 1) as a consequence of the spherical geometry of the corundum beads although not as much as predicted. Scanning electron micrographs of the surfaces of the beads showed that surface asperities were still present on the corundum beads (Figure 7) as a consequence of their method of formation, and so as the particles were not smooth but simply de-edged, the effect was not as pronounced as expected.

Theoretical models of various combinations of surface asperities and edges (Figure 8) were made in order to examine the results of these factors independently. It should be noted that these phenomena are produced only in spectral regions where the minerals are strongly absorbing.

Others have carried out experiments in the visible and ultra-violet regions and successfully adopted a theoretical treatment in which, because of their particle size to wavelength ratio, in addition to absorption, scattering by the surface asperities was a necessary term (Egan and Hilgeman, 1978).

Several improvements in our theory would be possible:
1) Our present approximate treatment (Emslie and Aronson, 1973) for the gross particle could be replaced by Mie theory.
2) The gross particle shape could be generalized to the case of a spheroid, in order to permit a much wider range of shapes, including needles and platelets, as well as chunky particles. This would be possible in view of the extension of the Mie theory to spheroids (Asano and Yamamoto, 1975).
3) A more accurate treatment of the effects of edges, especially for large particles, would consider an edge as a cylinder of macroscopic length but with cross-sectional dimensions less than the wavelength (Aronson and Emslie, 1979).
4) Scattering by the edges and asperities (Egan and Hilgeman, 1978) should also be included in the calculations.

In addition to these improvements it would be desirable to obtain

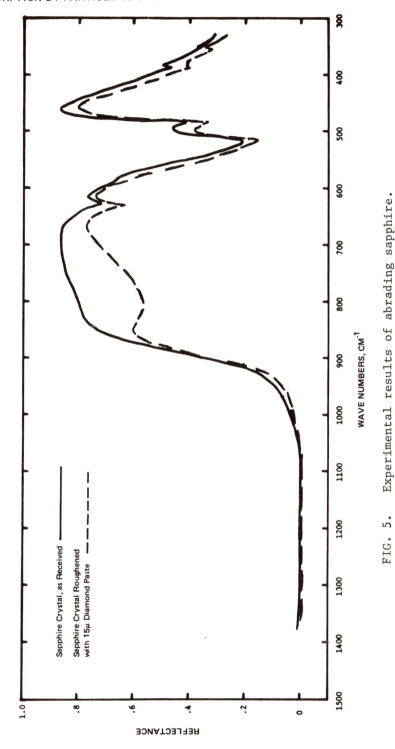

FIG. 5. Experimental results of abrading sapphire.

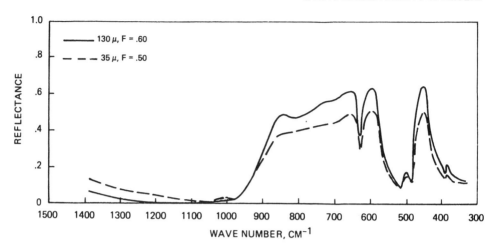

FIG. 6.   Experimental spectra of corundum beads.

a strictly theoretical proof of the enhanced absorption due to edges
for a simple geometry such as the cube. This might be possible by
means of the method of moments (Aronson *et al.*, 1975; Harrington
et al., 1972).

Comment (HUFFMAN):   Are you aware of the calculations of edge modes
in wedges, etc., done by Prof. Maradudin's group at U.C. Irvine?
These people are from a theoretical solid state physics background
and have not interacted strongly with light scatterers such as are
gathered here; however, I think his work might be quite pertinent to
your needs.

REFERENCES

Aronson, J. R., and Emslie, A. G., 1973, *Appl. Opt. 12*, 2573.

Aronson, J. R., and Emslie, A. G., 1975, *in* "Infrared and Raman
   Spectroscopy of Lunar and Terrestrial Minerals," C. Karr, Jr., Ed.,
   Academic Press, N. Y., p. 143.

Aronson, J. R., and Emslie, A. G., 1979, *Appl. Opt. 18*, 2622.

Aronson, J. R., Emslie, A. G., and Strong, P. F., 1975, Arthur D.
   Little, Inc., Interim Report to NOAA on Contract 03-4-022-121
   (April 1975).

Asano, S., and Yamamoto, G., 1975, *Appl. Opt. 14*, 29.

Egan, W. G., and Hilgeman, T., 1978, *Appl. Opt. 17*, 245.

Emslie, A. G., and Aronson, J. R., 1973, *Appl. Opt. 12*, 2563.

Harrington, R. F., Mautz, J. R., and Chang, Y., *Trans. IEEE, AP-20*,
   194.

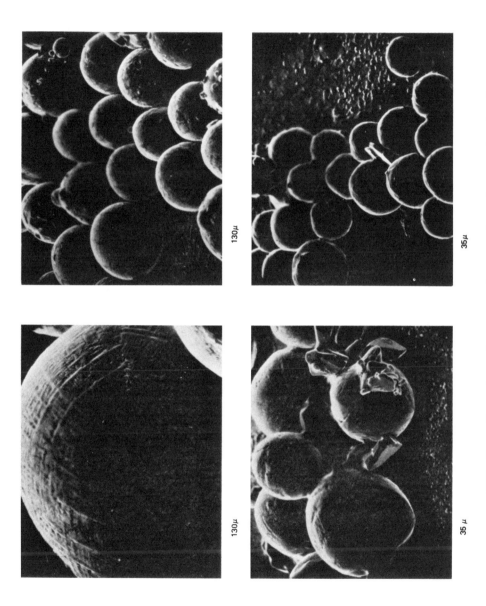

130μ

35μ

130μ

35 μ

FIG. 7. Scanning electron micrographs of corundum beads.

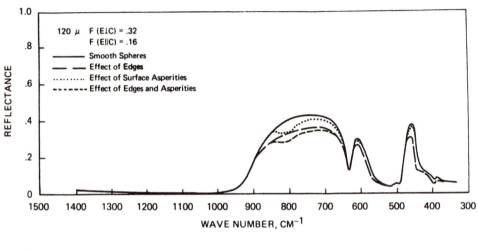

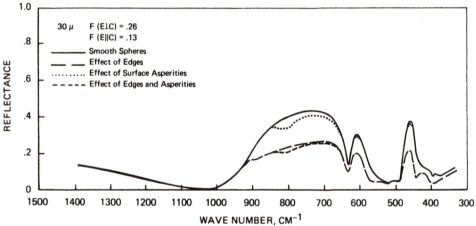

FIG. 8.   Theoretical spectra of corundum powders demonstrating the effects of edges and surface asperities. F denotes the volume fraction; E⊥C means the electric vector is perpendicular to the C axis.

ON THE SCATTERING FROM ARBITRARILY

SHAPED INHOMOGENEOUS PARTICLES - EXACT SOLUTION

C. Yeh and K. K. Mei

University of California, Los Angeles and EMtec

Los Angeles, California 90024

ABSTRACT

Discussion will be given on an exact approach to the problem of resonant scattering by arbitrarily shaped inhomogeneous particles. This exact approach is based on the Global-Local Finite Element which combines the contemporary finite element method with a hybrid-Ritz approach. Examples will be shown.

In order to understand the light scattering characteristics of naturally occurring particles, it has become increasingly clear that an accurate solution to the problem of light scattering by arbitrarily shaped inhomogeneous particles must be found. However, this is not an easy feat even with the help of high speed computers with very large memory capacities. The purpose of this paper is to introduce an exact approach called the Global-Local Finite Element Method which combines the contemporary finite element method with a hybrid-Ritz approach to yield the needed scattering solution.

To achieve a proper prospective on the state-of-the-art of various analytical/numerical techniques, a brief assessment will first be made.

(a) Geometrical theory of diffraction (Keller, 1962).

This method is mainly a high frequency technique which treats the smooth part of the scatterer by geometric optics and obtains the scattered fields from edges and glazing surfaces via solutions of canonical problems such as scattering by wedges, tips, etc. The

geometric theory of diffraction may yield reasonable results even at
resonant frequencies. However, it is still a technique most conven-
iently applied to convex and perfectly conducting targets at high
frequencies. When the targets are concave or transparent, tracing
the multiple scattering rays makes the method overly cumbersome to
use.

> Pros:
> > • Easy to learn
> > • Clear physical concept
>
> Cons:
> > • Difficult to apply to dielectric bodies
> > • Questionable validity at lower frequencies

(b) Method of moment (Harrington, 1968).

This method is based on the solution of an integral equation.
The equation is a surface integral equation if the scatterer is a
perfectly conducting body, a pair of coupled surface integral equation
if the scatterer is a homogeneous dielectric body, and a volume integral equation if the scatterer is an inhomogeneous dielectric body.
If the dielectric scatterer has a volume greater than $(0.1\lambda)^3$, the
demand of the method of moment for computer memory or time becomes
excessive.

> Pros:
> > • Efficient for thin conducting scatterer
>
> Cons:
> > • Limited to dielectric scatterer of small volume

(c) Perturbation method (Yeh, 1964).

This method is an extension of Mie scattering solution to
scatterers of non-spherical geometry by the boundary perturbation
technique. This technique is inherently limited to particle shapes
which are small perturbations of a sphere.

> Pros:
> > • Analytic expressions are available
>
> Cons:
> > • Uncertain convergence
> > • Limited to bodies that are small perturbations of
> >   a sphere
> > • Analytic expressions for higher-order terms are
> >   excessively complicated

(d) Point matching method (Oguchi, 1973).

Field expressions for the interior region of the scatterer and
for the exterior region are given in terms of complete sets of spherical harmonics. Boundary conditions are satisfied by point-matching
on the boundary surface.

Pros:
- Conceptually simple

Cons:
- Uncertain convergence
- Demands excessive computer time
- Limited to bodies that are close to spherical shapes

(e)   Extended boundary condition method (Barber and Yeh, 1975).

Integral representations for the fields can be derived which
satisfy the wave equation and all necessary boundary conditions. By
expanding the fields in terms of a complete set of spherical harmonics and by making use of analytic continuation techniques, one
may reduce the integral representations to a set of linear algebraic
equations.

Pros:
- May be used to treat scatterers of large volume and
  arbitrary shapes
- Numerical results can be generated very efficiently

Cons:
- Usually not applicable to inhomogeneous dielectric
  scatterers

According to the above discussion, none of the above methods can
be used to yield accurate results for scattering from resonant size
inhomogeneous bodies. This provides the impetus for us to search for
a new method to solve this problem. Initial indication shows that
the Global Local Finite Element Method is such a new method.

The term Global-Local Finite Element Method refers to numerical
analysis technique in which both the contemporary finite element and
classical Rayleigh-Ritz approximations are employed. This hybrid
Ritz method not only preserves the finite element modeling capability,
but adds the advantage of using *a priori* information regarding the
anticipated behavior to represent the total response in a given
problem. As a result, substantially fewer degrees of freedom are required in comparison to a problem using finite elements only. This
method was first suggested by Mote (1971, 1973) who demonstrated its
feasibility with beam and plate vibration examples.

The scattering problem is divided into two regions: the exterior
region and the interior region. Without any sacrifice of rigor, we
shall assume that the boundary surface between the exterior and interior regions is a sphere for three-dimensional problems and a circular cylinder for two-dimensional problems. It will further be
assumed that the boundary surface is so chosen that the medium in
the exterior region is homogeneous and that the inhomogeneous irregularly shaped absorbing object is contained within the interior
region. Solutions of wave equations in the exterior homogeneous
region are well-known. Therefore, knowing the interior fields at

the boundary surface will provide the necessary information to solve
the exterior fields by using the boundary conditions at the boundary
surface.   The finite-elements technique is used to find the interior
field.

To demonstrate our initial success of this technique, we shall
include here some of the preliminary results that we obtained for
the scattering characteristics of (a) a finite dielectric cylinder
with a spherical void and (b) two spheres.   Shown in Figure 1 is the
bistatic scattering cross section for a finite dielectric cylinder
of a diameter 1.87λ and length 1.87λ.   Plotted on the same figure

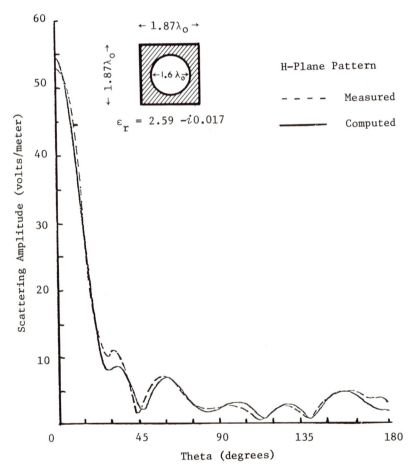

FIG. 1.   Bistatic scattering by a dielectric hollow finite cylinder.

are experimentally measured results.   Excellent agreement was found.
Another illustration showing the calculated and measured bistatic
scattering amplitudes for two neighboring conducting spheres is displayed in Figure 2.   The calculated results are obtained according
to the multiple scattering algorithm discussed earlier.   Again, very
good agreement is indicated.

The pros and cons of this Global Local Finite Element Method are
as follows:
   Pros:
        • Adaptable to scatterers of large volume arbitrary shapes
          and inhomogeneous material
        • Results are based on exact Maxwell's equations
        • Efficient computation
   Cons:
        • Requires extensive analytical and computational skill

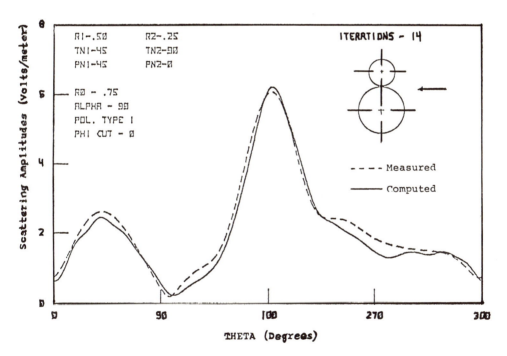

FIG. 2.   Bistatic scattering by two touching metallic spheres.

REFERENCES

Barber, P. and Yeh, C., 1975, *Appl. Opt.* 14, 2864.
Harrington, R. F., 1968, "Field Computation by Moment Method,"
  MacMillan, New York.

Keller, J. B., 1962, *J. Opt. Soc. Am.*, 52, 116.
Mote, C. D., 1971, *Int. J. for Numerical Methods in Eng.*, 3, 565.
Mote, C. D., 1973, *in* "The Mathematics of Finite Elements and
  Application," J. R. Whiteman, Ed., Academic Press, New York.
Oguchi, T., 1973, *Radio Science*, 8, 31.
Yeh, C., 1964, *Phys. Rev.*, 135, A1193.

LIGHT SCATTERING BY HEXAGONAL COLUMNS AND PLATES

Kuo-Nan Liou and Rich F. Coleman

University of Utah, Department of Meteorology

Salt Lake City, Utah 84112

ABSTRACT

Computations of light scattering by hexagonal columns and plates randomly oriented in two- and three-dimensional space are carried out by means of the geometrical ray tracing technique. Analyses are made for a number of crystal sizes, and for a visible wavelength of 0.55 μm and an infrared wavelength of 10.6 μm at which absorption plays a significant role. Scattering phase functions and degree of linear polarization computed from the ray tracing program are compared with those derived from experimental nephelometer measurements performed in the laboratory. Effects of the crystal size and shape on the scattering parameter will be described. We will also present the theoretical and experimental programs concerning light scattering by ice clouds to be carried out in the future.

1. INTRODUCTION

Cloud compositions change constantly and drastically with respect to time and space depending upon such variables as temperature, saturation ratio and atmospheric conditions. Perhaps there is no unique shape, size and orientation for the irregularly shaped ice crystals in the atmosphere.

Information on the scattering characteristics of non-spherical ice crystals is imperative to the development of remote sensing techniques for the cloud composition determination and to the understanding of the radiation budget of cirrus cloudy atmospheres.

In this paper we present some results of scattering and polarization calculations for hexagonal plates and columns by means of

geometrical ray tracing and present comparisons of these scattering
and polarization calculations with measured values obtained from the
laboratory laser scattering and cloud physics experiments.

2. RAY TRACING FOR HEXAGONAL CRYSTALS

The general geometry of light rays incident on a hexagonal
crystal is depicted in Figure 1. The geometry of the crystal is
defined by the length L and radius R, while the incident light rays
are described by the ray plane. Normal to the Z-axis, we define
the principle plane, which lies on the X-Y plane. A hexagon has six
equal sides and the top and bottom faces. To describe the geometry
of the hexagon with respect to the incident ray plane, seven variables
are required, *i.e.*, the length and radius of the hexagon and the
position of the principal plane, the position of the incident ray
on the ray axis, and three angles defining the orientation of the
crystal with respect to the incident ray, *i.e.*, the elevation angle θ,
the rotation angle Ψ and the azimuthal angle ϕ.

Having defined the variables involved, the ray tracing procedures may be outlined. We first find the position of the entry ray
in the (X,Y,Z) coordinates in terms of the seven geometrical variables. Through Snell's law, we find the refracted angles in terms
of the incident angle mapped on the principal plane and the elevation
angle. It follows that the position of the exit ray, the face that
the ray will hit and the geometrical path length in the crystal can
be determined through the procedures of the analytical geometry.
The procedures are then repeated for internally reflected rays.
Finally, we need to find the scattering angle with respect to the
incident ray, to perform the summation of the refracted and reflected
components and to carry out the normalization of the energy pattern
to get the scattering phase function.

The geometrical ray tracing equations for hexagons differ greatly
from those for spheres. Spheres have a curvature effect, whereas
hexagons do not. Also, a hexagon does not have the symmetry of the
geometrical path length that a sphere inherently possesses. Basically, the general equation for the scattered energy per unit angle
normalized with respect to the incident energy perpendicular to the
X-Z plane may be described by

$$E_{\ell,r}^{s}(\Theta,P) = \sum_{j} E_{\ell,r}^{s}(\Theta,P,\tau_{j}) \; e^{-2km_{i}\ell_{p}} \sum_{j} \cos\tau_{1j} \tag{1}$$

(equation (1) continues)

$$= \frac{e^{-2km_i \ell_p}}{\Sigma \ \cos\tau_{1j}} \begin{cases} \Sigma \ |r_{\ell,r}(\tau_{1j})|^2 \ \cos\tau_{1j}, \quad \text{P=0 (external reflection)} \\[2ex] \Sigma \ [1-|r_{\ell,r}(\tau_{1j})|^2][1-|r_{\ell,r}(\tau_{2j})|^2] \ \cos\tau_{1j}, \quad \begin{array}{l} \text{P=1 (two} \\ \text{refraction)} \end{array} \\[2ex] \Sigma \ [1-|r_{\ell,r}(\tau_{1j})|^2][1-|r_{\ell,r}(\tau_{(P-1)j})|^2] \\[1ex] \quad \times \prod_{n=2}^{P} |r_{\ell,r}(\tau_{nj})|^2 \ \cos\tau_{1j}, \quad \text{P≥2 (internal reflection)} \end{cases}$$

where subscripts ℓ and r denote the parallel and perpendicular polarization components, respectively, Θ the scattering angle, j the index for the entry rays, ℓ_p the ray path length in the crystal ($\ell_p=0$, when P=0), τ_1 the incident angle, which normally has three different values, p the index denoting the event of reflection and refraction and $\tau_2, \tau_3, \ldots$ are incident angles in the crystal. Since absorption is considered, absolute values need to be taken for the reflection and transmission components. Scattering energy patterns for two-dimensional and three-dimensional orientations may be subsequently computed by noting the specific relation of the incident angle and the elevation and azimuthal angles. For horizontal orientation cases, the incident angle is the elevation angle. But for general cases $\cos\tau = \cos\theta \ \cos\phi$.

The Snell's law governing the incident angles (θ,ϕ) and refracted angles (θ',ϕ') can be proven to be

$$m_r \ \sin \theta' = \sin \theta$$
$$m_r \ \frac{\cos \theta'}{\cos \theta} \ \sin \phi' = \sin \phi \tag{2}$$

where m_r is the real index of refraction.

To complete the ray tracing exercise, we have to include the diffraction pattern. The projection of a hexagonal column onto a horizontal plane clearly resembles a rectangle. The diffraction pattern for a rectangular aperture can be easily derived from Franhofer diffraction theory. It is given by

$$E^d(\Theta,\phi,L) = \frac{\sin^2(Rk \ \sin\Theta \ \cos\phi)}{(Rk \ \sin\Theta \ \cos\phi)^2} \ \frac{\sin^2((Lk/2) \ \sin\Theta \ \sin\phi)}{((Lk/2) \ \sin\Theta \ \sin\phi)^2} \tag{3}$$

where k is the wavenumber and R is the radius of the crystal. Clearly, three parameters are required to define the position of a hexagon in

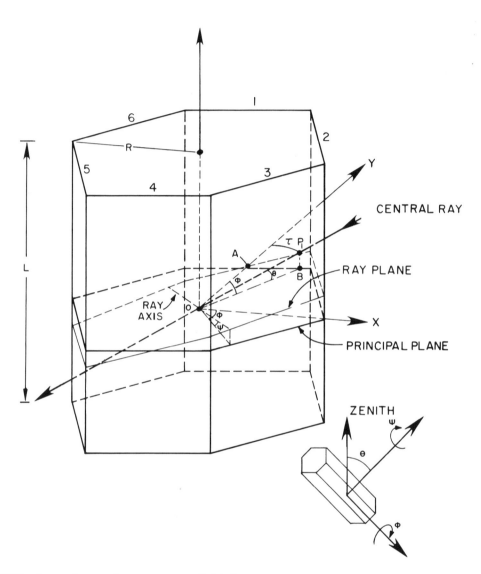

FIG. 1.   General geometry of light rays incident on a hexagonal
crystal.

reference to the incident ray, $i.e.$, the scattering angle Θ, the azimuthal angle ϕ and the geometrical length L. For horizontally oriented hexagons the diffraction pattern can be obtained by performing integration in ϕ from 0 to π, $i.e.$,

$$E_{2D}^d (\Theta,L) = \frac{1}{\pi} \int_o^\pi E^d (\Theta,\phi,L) \, d\phi \quad . \tag{4}$$

For three-dimensional random orientation, integration with respect to the length of hexagons is required. Thus, we find

$$E_{3D}^d (\Theta) = \begin{cases} \dfrac{1}{(1/2-R)} \displaystyle\int_R^{L/2} E_{2D}^d (\Theta,L') \, dL' \,, & \text{columns} \\[3ex] \dfrac{1}{(R-L/2)} \displaystyle\int_{L/2}^R E_{2D}^d (\Theta,L') \, dL' \,, & \text{plates .} \end{cases} \tag{5}$$

It should be noted that for plates, the approximate equation is less accurate because the major axis is on the plane of the hexagon.

Since the equations derived from the ray tracing procedure are in units of energy per degree, we must perform a normalization so that the scattering phase function can be derived. On the basis of the definition of gain with respect to isotropic scatterers, we find

$$\frac{G(\Theta) \ 2\pi \ \sin\Theta \ d\Theta \ r^2}{4\pi r^2} = E(\Theta) \ d\Theta \tag{6}$$

where r denotes the distance and the gain is normalized such that

$$\frac{1}{4\pi} \int_{4\pi} G(\Theta) \ \sin\Theta \ d\Theta d\phi = 1 \quad . \tag{7}$$

Thus, the gain is equivalent to the phase function commonly used in radiative transfer.

3. RESULTS AND DISCUSSIONS

Figure 2 shows the scattering patterns due to geometrical reflection and refraction for horizontally oriented and randomly oriented columns with lengths and radii of 300 and 60 μm, respectively, incident by a visible wavelength of 0.55 μm. The index P in the

diagram denotes the contribution of the scattering energy; P=0,
external reflection, P=1, two refractions, and P ≥ 2, internal reflection. The dashed and dashed-dot lines represent the scattering
patterns for horizontally oriented columns with elevation angles of
0° (normal incidence) and 42°, respectively, while the solid curve
denotes the scattering pattern for random orientation. The major
features for these three cases are the strong forward scattering and
halo in the region of 20°-30°. For horizontal orientation, the halo
feature shifts to a larger scattering angle when the incident angle
increases. We see a 8° difference for incident angles of 0° and 42°.
Owing to the shift of the halo features for different incident angles,
the halo feature in the case of random orientation is broadened and
smeared out. The less pronounced 46° halo features are also evident
in cases of random orientation and horizontal orientation with normal
incidence. The strong peak at 84° in the case of horizontal orientation with an incident angle of 42° is strictly due to the external
reflection. Note that the scattering pattern beyond 84° is caused
by the end effects and internal reflections. For random orientation
and horizontal orientation with normal incidence, the backscattering
is primarily produced by one internal reflection. The less pronounced backscattering in random orientation case is the result of
the averaging over many oblique incidence cases.

Shown in Figure 3 are the normalized scattering phase functions,
which include both the diffraction and geometrical reflection and refraction contributions, for horizontally oriented plates with parallel
incidence for two wavelengths of 0.55 and 10.6 µm. (The random orientation pattern for plates will be discussed in the next paragraph.)
At the wavelength of 10.6 µm, the real and imaginary parts of the
refractive index are, respectively, 1.097 and 0.134. Clearly, absorption must be significant. Comparison with a real refractive index of
1.31 for the 0.55 µm wavelength shows that from about 50 to 180° the
scattering is much reduced, especially in the backscattering directions where one internal reflection, which is the dominant contri-
butor to the scattering process, is decreased due to significant absorption. At the 10.6 µm wavelength, plates also produce a halo
feature at 7° but its magnitude is on the same order as the forward
diffraction value. Thus, it does not become differentiable from the
forward diffraction peak. The diffraction peak for 10.6 µm is much
broadened and its maximum intensity is much smaller than that for
0.55 µm as shown in Figure 3.

Comparison of the scattering phase functions for randomly oriented columns with lengths of 300 µm and radii of 60 µm and plates with
radii of 125 µm and lengths 25 µm is illustrated in Figure 4. The
most significant scattering differences between plates and columns
are the much lower forward peak for plates and the much lower side
scattering for columns. The 22° and the less pronounced 46° halos
for columns are relatively stronger than those for plates. Both
scattering patterns depict the very narrow diffraction peak, strong

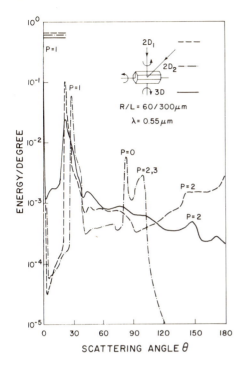

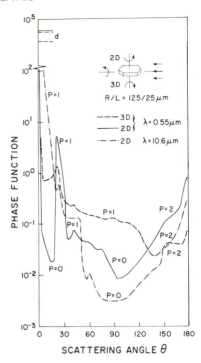

FIG. 2.   Scattered energy per
degree for randomly oriented
columns (3D) and for horizontally
oriented columns (2D) with elevation angles θ of 0° and 42°.

FIG. 3.   Phase functions for
randomly oriented plates (3D) and
for horizontally oriented plates
(2D) with elevation angles θ of
0° and 42°.

22° halo feature and broad peak at about 150°. The shift in the
150° peak from columns to plates appear to be due to the fact that
randomly oriented columns are much more like a spheroid than plates.

     The degree of linear polarization is shown in Figure 5 for
randomly oriented columns and plates as well as for spheres based on
the geometrical ray tracing for comparison purposes. The polarization pattern for plates remains negative from 0° to about 66°,
whereas for columns negative polarization only extends from 0° to
about 39°. The strong polarization maximum for plates at about 136°
is caused by external reflection (p = 0). Such a maximum occurs at
about 70° for columns. The positive polarization peaks at about
156° and 178° for columns associated with one internal reflection
(p = 2). There is a slight negative polarization for plates in the
backscattering direction from about 165° to 180°. The polarization
patterns for non-spherical plates and columns differ significantly
from the polarization produced by spheres. Large spheres (*i.e.*, in

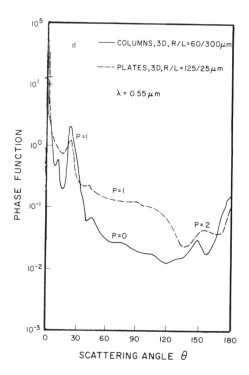

FIG. 4.   Phase functions for randomly oriented (3D) columns and plates.

geometrical regions) generate strong polarization at about 80° due to the external reflection and at the first (~138°) and second (~126°) rainbow angles caused by one and two internal reflections, respectively. The apparent and significant differences in polarization patterns caused by the shape factor may provide a practical and feasible means for the identification of spheres, plate-like and column-like particles in clouds. Also shown in this figure is the polarization pattern derived from experimental data for plates having a modal diameter of 20 μm. There is general agreement between the measured and calculated polarization patterns for plates despite the size difference.

Figure 6 shows comparisons of the normalized phase functions for plates derived from a number of nephelomemeter measurements in the laboratory (Sassen and Liou, 1979) and from ray tracing calculations. Three samples of plates with modal diameters of 1.5, 3.5 and 20 μm were produced in the cold chamber during the scattering experiments. Experimental values reveal that the side scattering for plates increases when the particle size increases. This increase is contrary to the scattering behavior of spherical water drops whose side scattering in the normalized scattering phase function generally reduces with increasing sizes. The much larger side scattering from ray optics calculations is physically understandable in view of the large plates considered. Figure 7 illustrates another comparison of the normalized phase function for clouds composed of a mixture of small plates and columns obtained from three scattering measurements with that derived from ray tracing calculations for columns. Again, the ray optics results show much stronger side scattering and a pronounced 22° halo as well as a less noted 46° halo, which small columns and plates are unable to produce. Note that the forward scattering for small columns and plates is greater than that of large columns in the regions from about 10° to 20°.

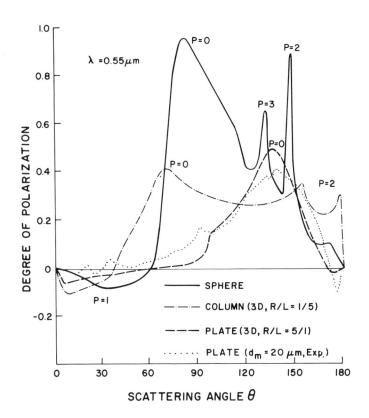

FIG. 5.   The degree of linear polarization for randomly oriented
columns and plates and for spheres. Also shown is the polarization
pattern derived from experimental data for plates having a modal
diameter of 20 μm.

        It would be desirable to generate larger plates and columns in
the cloud chamber so that more significant comparisons between ˙
measured scattering data and ray optics calculations could be carried
out to cross-check the experimental and theoretical results more

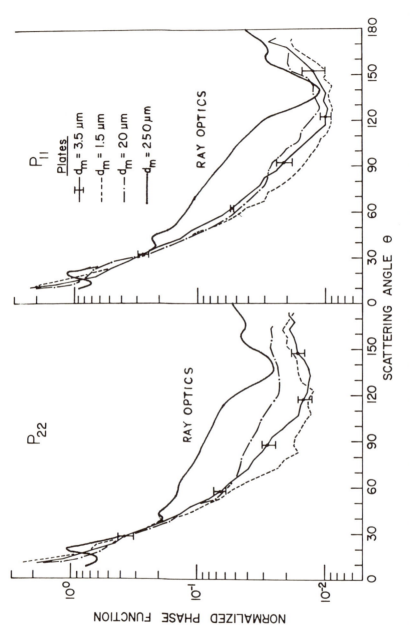

FIG. 6. Normalized phase function for plates from a number of nephelometer measurements in the laboratory and from ray tracing calculations.

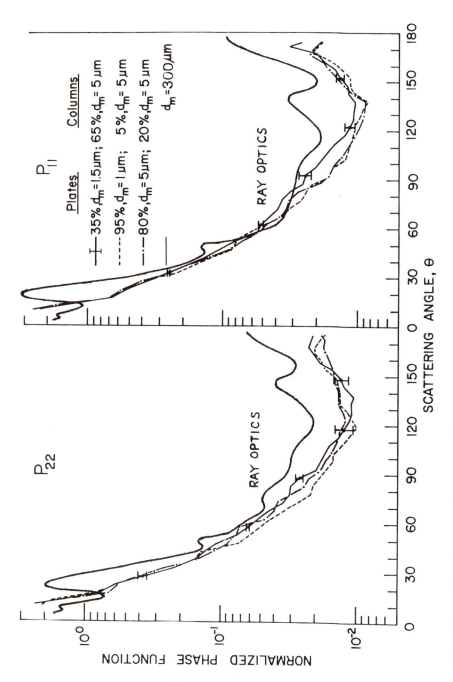

FIG. 7. Normalized phase function for mixtures of plates and columns from a number of nephelometer measurements in the laboratory and for columns from ray tracing calculations.

reliably and comprehensively. Finally, it should be noted that information and physical understanding of the basic scattering parameters for oriented columns and plates are required to perform radiative transfer calculations for cirrus clouds and to develop active remote sensing techniques for the identification of the phase, shape and size of cloud particles.

Acknowledgements. This research was supported by the Meteorology Program, Division of Atmospheric Sciences, National Science Foundation under Grant ATM78-26259.

Questions (JAYAWEERA): In natural clouds, the basic shapes you described are more the exception than the rule. To what extent could you apply the ray tracing technique to derive the scattered intensities from natural ice crystals? Also, would it be more prudent to look to an empirical relation for scattering functions of natural ice crystals because the ray tracing for these crystals will be extremely time consuming and laborious?

Answer (LIOU):   In answer to your questions, I would like to point out that the fact that we see halo phenomena in the atmosphere indicates that a large number of ice crystals have basic hexagonal column and plate structures. To understand physically the scattering characteristics of ice crystals, it is necessary, but not sufficient, to perform theoretical analyses and computations in terms of the ray tracing technique. As you have correctly pointed out, naturally occurring ice crystals have in general more complex shapes than these simple columns and plates. It is, therefore, extremely important to carry out controlled laboratory light scattering experiments in which known sizes and shapes and perhaps orientations may be generated and to make use of the experimental results to assess the theoretical computations. I have shown in my presentation some comparisons between results derived from ray tracing computations and laboratory scattering measurements. Moreover, I also feel that it is significant to conduct field scattering experiments involving ice crystals so that quantitative scattering properties of natural ice crystals can be obtained under a variety of atmospheric conditions. However, without theoretical analyses and computations, it is not possible to understand the physical significance of particular scattering features and general properties of irregularly shaped ice crystals in conjunction with remote sensing exploration and radiative transfer studies.

REFERENCE

Sassen, K. and Liou, K. N., 1979, *J. Atmos. Sci.*, *36*, 838-851.

SCATTERING OF RADIATION BY A LARGE PARTICLE WITH A ROUGH SURFACE

Sonoyo Mukai and Tadashi Mukai

Kanazawa Institute of Technology

Ishikawa, Japan

Richard H. Giese

Ruhr Univ., Bereich Extraterrestrische Physik

Bochum, FRG

ABSTRACT

The scattering of radiation by a large particle bounded by a rough surface is treated. The surface roughness is represented by a slope distribution function. Since surface roughness causes multiple reflections of incident light, multiple scattering theory is applied to obtain the reflected intensity. It is shown that experimental results are well explained by this model.

1. INTRODUCTION

According to Cox and Munk (1954), sea waves play a significant role in sunlight reflection by the ocean. That is, there is an evident distinction between the angular distribution of light reflected by a wind-ruffled sea surface and that reflected by a mirror-like calm ocean. This means that light reflected by a rough surface is distributed around the specular point, while the light is specularly reflected alone if the surface is completely smooth (Mukai, 1977).

In the scattering of light by a particle with a smooth surface, only once reflected light contributes to the reflectance. On the

contrary, a rough-surfaced particle causes multiple reflections of
incident light just as sea waves do. This fact, namely that surface
roughness disperses the reflected light, motivates us to examine the
scattering by a large particle with a rough surface. Several authors
have discussed the problem of scattering by irregularly shaped particles (see references in our recent paper, 1979). Here the irregularity of the particle is represented by surface roughness, while the
interior of the particle is considered to be homogeneous. Furthermore, we assume that this irregularity is sufficiently small compared
to the particle size and that the surface consists of a discrete set
of surface elements with proper slopes whose probability of occurrence
is described by a slope distribution function.

Since the successive reflection of light by such surface elements
can be treated stochastically, it is shown here that multiple scattering theory can predict the intensity of multiple reflected light by
the rough surface. The numerical results of scattering properties
derived from the above procedure give good agreement with experimental results as is shown in the final section.

2. METHOD

When particles are larger than the wavelength of radiation, the
scattering may be treated as an approximate combination of diffraction, geometrical reflectance, and transmission. For the sake of
simplicity we restrict our discussions to absorbing particles; therefore, we can neglect the component of transmission. Since the diffracted light is not affected by the trivial roughness of the particle
surface, the diffracted intensity is easily calculated by using well-
known formulae. Then, let us discuss the computation of intensity
due to external reflection by a rough surface.

Reflection of incident light from the surface is based on
Fresnel's law which depends on the angle of incidence with respect
to the normal of the reflecting surface element. Since the surface
element has a slope which is expressed by a slope distribution
function, the intensity and direction of reflected light are functions
of the incident angle governed by this slope distribution. In other
words, a reflecting surface element behaves as a grain with an anisotropic phase function determined from both Fresnel's reflection law
and the slope distribution function. We can say, therefore, that the
intensity of multiple reflected light by the rough surface is obtained by solving the radiative transfer problem for a medium which
consists of grains.

The equation for this transfer problem takes the following form

$$\mu\left(\partial I(\tau,\Omega)/\partial\tau\right) = I(\tau,\Omega) - (1/4\pi)\int \tilde{P}(\Omega,\Omega')I(\tau,\Omega')d\Omega' , \qquad (1)$$

where I is the intensity of radiation represented by one-column vector
of Stokes parameters (Chandrasekhar, 1950). P is the phase function,
and the solid angle Ω is as a function of μ and ϕ; μ is the cosine of
the angle between the outward normal and the direction of light and
ϕ is the azimuthal angle. The optical depth τ is the interaction
length of light and grains. Note that, as mentioned before, these
grains correspond to the surface elements of the particle. Since the
number of reflections suffered near the rough surface are considerable, we shall treat the transfer equation in a medium of semi-
infinite optical thickness.

Apart from the transfer equation, we discuss the slope distribution for rough surface. Certainly an arbitrary slope distribution
can be adopted for the present calculation. The slope distribution
of the sea waves, however, is approximately described by a Gaussian
distribution (Cox and Munk, 1954), and it is also reported that the
averaged roughness of a metallic surface may be similarly represented (Pernick, 1979). So we take the isotropic Gaussian distribution
as an example of a slope distribution for the rough surface. The
mean square slope of the Gaussian distribution function, σ^2, characterizes the roughness of the surface. This implies that roughness
increases as σ^2 increases, a smooth surface like a mirror is expressed by $\sigma^2=0$, and a completely random rough surface corresponds
to $\sigma^2\to\infty$.

The phase matrix $\tilde{P}$ in equation (1) has to be determined in a
way that represents the reflection pattern by a rough surface. For
simplicity, the reflecting diagram resulting from the incident light
on a surface with slope β for the azimuth independent case is shown
in Figure 1. The angles θ_o and θ are polar angles of incident and
reflected light, respectively, and ω is Fresnel's reflection angle.
It is readily seen that the surface slope β and Fresnel's angle ω
are determined from fixed values of θ_o and θ. Once we get the angles
β and ω, we can calculate the values of the slope distribution

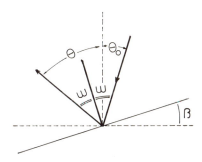

FIG.1.   Diagram showing the reflection of light by a surface with
slope β.

function and Fresnel's reflection coefficient, and then we get the phase matrix $\tilde{P}$ as the product of these two values.

The equation of transfer (1) is solved by the method of successive orders of scattering (Uesugi and Irvine, 1970), and then the numerical results are integrated over the particle disk (Horak, 1950) to compare with the experimental results. A detailed treatment is presented in Mukai and Mukai (1979).

3. COMPARISON WITH EXPERIMENTS

Let us introduce first the results of an infinitely long circular cylinder (Mukai *et al.*, 1979). In Figures 2a and 2b we compare numerical results to experimental ones by Andorf (1978) for a tungsten cylinder with size parameter $\alpha=79.4$ and refractive index $m=3.46-0.25i$. The solid curves represent the experimental results. Open circles and dots denote the numerical results of relative intensities that are normalized at a scattering angle of $\Theta=75°$ for $\sigma^2=0$ (smooth surface) and $\sigma^2=\infty$ (completely random rough surface, which means all slopes occur with equal probability), respectively. From this comparison with the experiments we get the following conclusions:
   1) It is impossible to explain the backward scattering feature with a smooth surface cylinder.
   2) As roughness increases, *i.e.* as σ^2 increases, a "hump" appears in the intensity at large scattering angles.
   3) So far as the normal component (in Fig. 2a) is concerned, a random rough surface cylinder shows fairly good agreement with the experimental results.

Next, Figure 3 shows the scattered intensity due to a spherical particle with a completely random rough surface and with $\alpha=31.2$ and $m=1.65-0.25i$. Dots represent the experimental results for a compact irregular particle, and the dashed curve denotes the numerical values of a smooth-surface sphere which are reproduced from Giese *et al.* (1978). Numerical results obtained here are represented by a solid curve. We can say that this model is well suited for explaining the measured intensity for a compact irregular particle, while Mie calculations of a smooth surface sphere do not coincide with the experimental results, especially in the backward part.

Our model will be checked when we know the rigorous slope distribution of the sample surface in the experiments. Cases for more general types of irregular particle, *e.g.* fluffy particles, will be shown in a later work.

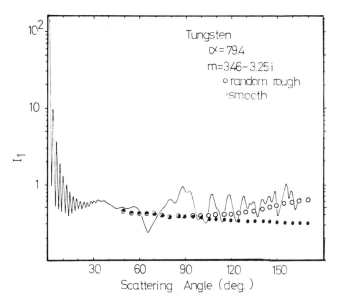

FIG. 2a.   Normal component of the relative intensity for a tungsten
cylinder, where α is the size parameter and m is the refractive index.
The experimental results by Andorf (1978) are shown by the solid
curve.

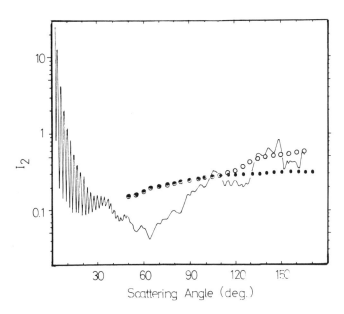

FIG. 2b.   Parallel component.   Same as Fig. 2a.

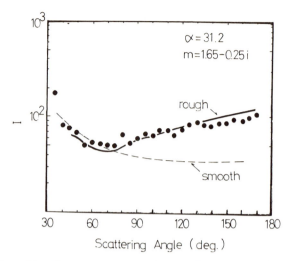

FIG. 3.   Intensity for a compact irregular particle. Dots represent
the measurements by Giese *et al*. (1978). Since the diffraction
pattern dominates below the scattering angle 30°, the intensity
curve in this region is omitted.

Question (CUZZI): How significant, both in number and total contribution, are higher order reflections compared to single order reflection?

Answer (MUKAI):   The contribution of higher order reflections
depends on the albedo for single reflection in general. In our
calculations we consider absorbing particles, so single reflection
is dominant for intensity. However, as far as the degree of polarization is concerned, higher order reflections have a significant
effect even for absorbing particles.

Questions (SHETTLE): I have 2 questions: First, in the comparison
between your calculations and the measured angular scattering from a
rough cylinder, the measurements show a number of features as a
function of scattering angle which do not appear in your calculations.
Why? Could this be due to regularities or periodicities in the roughness of the cylinder, since your calculations assume a random distribution of roughness? Second, are you making any assumptions on the
characteristic dimensions of the surface roughness relative to the
wavelength?

Answers (MUKAI): 1) Certainly fluctuation of measurements may be
caused by periodicities in the surface roughness. A rigorous check
of our model needs the correct knowledge of slope distribution of the
sample surface. Unfortunately, since there are only a few experiments, we can not discuss this in detail. 2) You can imagine the
irregularities on the particle to be just like the sea waves. Sea
waves are large compared to the wavelength of radiation; on the other
hand, they are small compared to the ocean.

REFERENCES

Andorf, F., 1978, Staatsexamensarbeit, Ruhr-Universitat Bochum.

Chandrasekhar, S., 1950, "Radiative Transfer," Oxford Univ. Press, London, 43.

Cox, C. and Munk, W., 1954, *J. Opt. Soc. Amer.* 44, 838.

Giese, R. H., Weiss, K., Zerull, R. H. and Ono, T., 1978, *Astron. & Astrophys.* 65, 265.

Horak, H. G., 1950, *Astrophys. J.* 112, 445.

Mukai, S., 1977, *Astrophys. & Space Sci.* 51, 165.

Mukai, S. and Mukai, T., 1979, *Icarus* 38, 90.

Mukai, S., Mukai, T. and Giese, R. H., 1979, submitted to *Appl. Opt.*

Pernick, B. J., 1979, *Appl. Opt.* 18, 796.

Uesugi, A. and Irvine, W. M., 1970, *Astrophys. J.* 159, 127.

THE MICROWAVE ANALOG FACILITY AT SUNYA:

CAPABILITIES AND CURRENT PROGRAMS

Donald W. Schuerman

State Univ. of N.Y. at Albany, Space Astronomy Lab

Albany, New York 12203

ABSTRACT

This newly renovated facility at the Space Astronomy Laboratory (SUNYA) is a powerful experimental tool for investigating scattering by non-spherical particles. Single particle extinction data, averages of the scattering function over random particle orientation, precise knowledge of the size, shape, index of refraction, and particle-shape versus incident-beam geometry are advantages not realized by other experimental methods. Three on-going programs at the facility are also briefly described.

1. INTRODUCTION

Experimental methods of investigating resonant light scattering by single particles can be divided into two categories - optical and analog techniques. In the former method, empirical results for a specific particle can be obtained at the expense of sacrificing exact knowledge of the physical inputs to the scattering problem. In particular, the particle-shape versus incident-beam geometry is never known for an arbitrary particle, and even the determination of the particle shape requires a facility capable of handling individual submicron particles. However, optical methods are well suited for making "in situ" measurements and for obtaining "real world" phase functions which can immediately be employed in problems of radiative transfer. The analog method, on the other hand, provides a precise description of the physical inputs to the scattering problem. The size, shape, and index of refraction of the target particle are known to an arbitrary accuracy. Of greatest importance is the fact that the particle orientation with respect to the incident beam is

227

not only known at all times during the measuring process, but the
particle orientation can even be programmed beforehand. These assets
make the microwave analog technique ideally suited for studying the
physics of the scattering process. The price paid here is that the
scattering target may not be isomorphic to a real world particle.

2. THE MICROWAVE ANALOG TECHNIQUE

The microwave scattering analog relies on the principle of
electromagnetic similitude (Stratton, 1941), an exact relationship
based on Maxwell's equations, which states that the physics of the
scattering process depends only upon the ratio of particle size to
wavelength. In the microwave facility at the Space Astronomy
Laboratory, a collimated beam 25 cm in diameter and of wavelength
3.18 cm is scattered by an arbitrary target to a parabolic receiver
located at any scattering angle such that $5° < \theta < 175°$ or at $\theta = 0°$
or 180°. In the angular scattering mode, the first four components
of the intensity matrix (F_{11}, F_{12}, F_{21}, F_{22}, using the rotation of van
de Hulst, 1957) are measured. In the forward scattering mode ($\theta=0°$)
both the amplitude and phase of the amplitude matrix $S(0)$ are
measured so that, via the optical theorem, the extinction cross
section can be obtained. In the back scattering mode ($\theta=180°$), just
the F_{11} and F_{22} components can be determined since the same continuous wave dipole antenna is used for transmitting and receiving. In
both the angular and forward scattering modes, a unique design of
continuous-wave cancellation allows the radiation in the forward
scattering cone of the target to be separated from the background
radiation of the transmitter so that small scattering angles can be
investigated. This is not possible for optical methods. Some
specific examples of our microwave analog extinction measurements
are presented in the article by Ru T. Wang in this volume. Details
of the microwave plumbing and associated electronics can be found
in the article by Lind *et al.* (1965) and the report of Wang and
Greenberg (1978).

Scattering targets of arbitrary shape are machined or molded
from materials that have, in the microwave region, the same properties (dielectric constant and conductivity) as the corresponding
"real world" particles have in the visible wavelength region. By
controlling the density of soft expandable plastics like polystyrene,
any of a wide variety of indexes of refraction can be duplicated.
The absorptivity is controlled by the proper admixing of conducting
inclusions such as carbon. The resulting real and imaginary parts
of the index are measured via the slotted waveguide method of Roberts
and van Hippel (1946).

The target, in operation, is suspended in the incident beam by
nylon monofilament lines which are sufficiently thin to cause negligible scatter. A computer controlled rotating table and "tilt" mechanism

provide the means to automatically orient ("point") the target in
any of 4π directions. This recently automated orientation mechanism
allows on-line averages of scattering data to be taken over particle
orientation. Our standard measuring procedure is to average particle
scattering data taken at 272 particle orientations (or "pointing"
directions) which are (almost) uniformly spaced over the sphere; the
272 individual measurements are maintained on magnetic tape for any
further investigations requiring specific particle orientations.
This capability to perform an *exact* average over particle orientation
is extremely important. Scattering data which has been averaged in
the appropriate manner corresponds to the scattering from a random
cloud of particles. Other possibilities can also be realized such
as the scattering from a cloud of particles aligned according to
a given distribution representing, say, charged, elongated particles
in a magnetic field. Such experiments relate to scattering in the
interstellar and interplanetary media.

In Table 1, the capabilities of the Microwave Analog Facility
at SUNYA are summarized. The time required to measure the complete
scattering from a single particle - *i.e.*, to obtain the extinction
cross section and the angular intensities $F_{11}(\theta)$, $F_{12}(\theta)$, $F_{22}(\theta)$,
all averaged over particle orientation, - is about two days. A view
of the facility looking downstream from the transmitter is shown in
Figure 1.

TABLE 1

SUMMARY OF CAPABILITIES AT THE MICROWAVE
ANALOG FACILITY, SPACE ASTRONOMY LAB, SUNYA

Measured quantities :	$S_1(0)$, $S_2(0)$, $S_3(0)$, $S_4(0)$
(van de Hulst's notation)	both amplitude and phase.
	$F_{11}(\theta)$, $F_{12}(\theta)$, $F_{21}(\theta)$, $F_{22}(\theta)$ for any θ such that $5° < \theta < 175°$.
	$F_{11}(\pi)$, $F_{22}(\pi)$
Derived cross-sections :	C_{ext}, $C_{scatt}(\theta)$, C_{abs}, $C_{rad.\ press.}$
Range of size parameter:	$1 < \pi d/\lambda < 20$ where d is the largest particle dimension.
Range of indexes :	$m = m' - im''$, $1.07 < m' < 2.0$; $.002 < m'' < 0.5$ and $m = \infty$.
Particle Orientation :	Measured quantities can be averaged over random or any other well-defined distribution of particle orientations.

FIG. 1.   The Microwave Analog Facility.   The angular scattering receiver is here positioned to the left of the forward scattering (black) receiver.   (The transmitter is not shown).   The orientation mechanism (center foreground) controls the orientation of the target which, in this case, is two spheres in contact.   The bank of electronic gear behind Dr. Ru T. Wang is used to measure both the amplitude and phase of the forward amplitude matrix.

3.   CURRENT PROGRAMS

On-going programs at the Microwave Scattering Facility include:

(1)   The comparison of computed to measured values of the scattering by arbitrary spheroids. A computer code recently developed by Richard Schaefer of SUNYA expands the method of Asano and Yamamoto (1975) to include complex indexes. Experimental results at specific particle orientations are used as a check on the computer program. Conversely, this code is used to calibrate the measured, off-diagonal components of the F matrix. Mie theory is not useful for this task because these components vanish for spheres.

(2)   Measurements of scattering due to particles which we call "bird's nests". These particles, which simulate an aggregate of many small, silicate cylinders held together in an ice matrix, seem to best represent the scattering observed from interstellar and interplanetary dust (Greenberg, 1980).

(3)   A systematic study of the effects of particle shape on scattering. This program, funded by the U.S. Army Research Office, is designed to identify and categorize any signatures in the scattering functions which can be attributed to simple shapes and/or particle dimensions. The complete set of scattering measurements (see Table 1), averaged over random particle orientation, is being obtained for the following shapes: finite cylinder (4 to 1 aspect), prolate spheroid (4 to 1), prolate spheroid (2 to 1), sphere (for calibration), oblate spheroid (2 to 1), oblate spheroid (4 to 1), and a disk (4 to 1). This group of measurements from seven particle shapes will be repeated for four different sizes. In all, 28 well-defined particles ($m \simeq 1.6$) will be used in this attempt to systematically define the effect of particle shape, particularly when averages are taken over particle orientation.

Acknowledgement. The author acknowledges the support of the U.S. Army Research Office under contract number DAAG-29-78-G0024.

REFERENCES

Asano, S., and Yamamoto, G., 1975, *Appl. Opt.* *14*, 29.
Greenberg, J. M., 1980, From Interstellar Dust to Comets to the
   Zodiacal Light, to appear in "Solid Particles in the Solar System,
   Proceedings of IAU Symposium No. 90," D. Reidel Publ. Co., Dordrecht.
Lind, A. C., Wang, R. T., and Greenberg, J. M., 1965, *Appl. Opt.* *4*,
   1555.
Roberts, S., and von Hippel, A., 1946, *J. Appl. Phys.* *17*, 610.
Stratton, J. A., 1941, "Electromagnetic Theory," McGraw Hill, New
   York.

van de Hulst, H. C., 1957, "Light Scattering by Small Particles,"
  Wiley, New York.
Wang, R. T., 1980, Extinction Signatures of Non-spherical/Non-
  isotropic Particles, this volume.
Wang, R. T., and Greenberg, J. M., 1978, Final report NASA NSG 7353
  (Aug., 1978).

OBSERVATION OF LIGHT SCATTERING FROM ORIENTED

NON-SPHERICAL PARTICLES USING OPTICAL LEVITATION

A. Ashkin and J. M. Dziedzic

Bell Telephone Laboratories

Holmdel, New Jersey   07733

ABSTRACT

The optical levitation technique has been used to assemble
spheroids and other non-spherical particles of simple shape. These
particles are held at fixed orientation in laser beams where their
light scattering properties can be studied.

We discuss here a new extension of the optical levitation technique to the study of light scattering from non-spherical or irregular
particles. The basic levitation technique (Ashkin, 1970; Ashkin and
Dziedzic, 1971) involves use of the forces of radiation pressure to
manipulate and stably trap single dielectric particles in the approximate range of 1-100 microns. In the past, experiments have been
performed in various media such as liquids (Ashkin, 1970), air
(Ashkin and Dziedzic, 1971, 1974, 1975), and vacuum (Ashkin and
Dziedzic, 1976a, 1977a). The technique is useful wherever small
particles play a role. Applications in the fields of atmospheric
physics (Ashkin and Dziedzic, 1975), laser fusion (Ashkin and Dziedzic,
1974, 1977a), and photoemission (Ashkin and Dziedzic, 1976b) have
been discussed. The levitation technique is, however, most clearly
useful in the field of light scattering. In fact, the basic radiation
pressure forces used in levitation are directly derived from a
knowledge of the light scattering from a particle. Recently a study
of the wavelength dependence of the levitating force on highly
perfect liquid drops led to the first observation of ultra-narrow
surface wave resonances (Ashkin and Dziedzic, 1977b). These narrow
resonances permit one to measure the relative sizes of spheres to
an accuracy of about 1 part in 10^5. Absolute measurement of the

size of spheres to this accuracy is possible using theoretical Mie
calculations of such surface-wave resonances (Chýlek *et al.*, 1978).
The paper by P. Chýlek, A. Ashkin and J. Kiehl in this volume, in
part, discusses observations of surface wave resonances in the angular distribution of the scattered light. Use of surface wave
resonances probably represents the most accurate determination of
particle size by light scattering techniques. One can, of course,
use the narrow surface wave resonances as probes of small distortions
of the perfect sphere.

In this work we concentrated on those non-spherical particles
that can be assembled from spheres, such as spheroids, teardrops,
doublet, triplet, and quadruplet particles, and various spheres
within spheres. These particles were assembled in air using two
independently adjustable laser beams to hold and combine different
spheres. By combining solid quartz spheres we made the doublet,
triplet and quadruplet particles. By colliding a doublet consisting
of two equal quartz spheres with a drop of index matching oil, one
finds that the liquid preferentially fills in all the crevices of
the doublet by surface tension and one ends up with a particle which
is a very close approximation to an optically homogeneous spheroid.
If the initial spheres of the doublet are unequal, one gets a teardrop
shaped particle. If one of the spheres is quartz and the other is
a liquid drop of different index of refraction, we end up with a
sphere within a sphere. We found that once such a particle is formed,
it remains levitated with a fixed orientation in the beam where its
size, shape, and scattering properties can be studied in detail.
The size, shape and near field light scattering properties are
determined by focusing microscopes on the particle and projecting
the enlarged images directly on a viewing screen. Far field
scattering can be observed by simply placing viewing screens in the
far field with all microscopes removed. This permits us to directly
photograph the entire scattering pattern from oriented particles free
of any interfering obstacles. By following the evolution of the
scattering patterns from a sequence of particles, starting for
example with a sphere and proceeding to a doublet, then to a
spheroid with an internal doublet of different index of refraction
and finally to an optically homogeneous spheroid, one is able to
acquire an intuitive grasp of the correspondence between features
of the particle and features of the near and far field scattering.

We believe that the new optical levitation techniques which we
have developed will make possible scattering measurements on each
of these new particles which are as detailed as those on spheres.
We further think that optical levitation will prove to be the method
of choice for studying light scattering from any of the particles
which fall in the range of this technique.

REFERENCES

Ashkin, A., 1970, *Phys. Rev. Letters* _24_, 156.
Ashkin, A. and Dziedzic, J. M., 1971, *Appl. Phys. Letters* _19_, 283.
Ashkin, A. and Dziedzic, J. M., 1974, *Appl. Phys. Letters* _24_, 586.
Ashkin, A. and Dziedzic, J. M., 1975, *Science* _187_, 1073.
Ashkin, A. and Dziedzic, J. M., 1976a, *Appl. Phys. Letters* _28_, 333.
Ashkin, A. and Dziedzic, J. M., 1976b, *Phys. Rev. Letters* _36_, 267.
Ashkin, A. and Dziedzic, J. M., 1977a, *Appl. Phys. Letters* _30_, 202.
Ashkin, A. and Dziedzic, J. M., 1977b, *Phys. Rev. Letters* _38_, 1351.
Chýlek, P., Ashkin, A. and Kiehl, J., 1980, "The Role and the
    Description of Surface Waves in Light Scattering," this volume.
Chýlek, P., Kiehl, J. T., and Ko, M. K. W., 1978, *Phys. Rev. A* _18_,
    2229.

THE EFFECT OF AN ELECTRIC FIELD ON THE BACKSCATTERED

RADIANCE OF A SINGLE WATER DROPLET

M. J. Saunders

Bell Laboratories

Norcross, Georgia   30071

ABSTRACT

Microscopic droplets formed by the condensation of water vapor
on sodium chloride crystals are supported on spider threads and subjected to a horizontal electric field. The shape of the droplet and
the near field surface wave radiance are dependent upon the intensity
of the field. The spider thread method of support should permit the
angular scattering pattern of an isolated irregularly shaped particle
to be obtained.

Scattering of electromagnetic radiation by non-spherical
particles is receiving increased attention by the scientific
community, and experimental approaches to the problem involve both
scattering from a cloud of particles and scattering from a single
particle. The major problem in the single particle scattering
approach is the method whereby the single particle is supported, for
the support mechanism must be one that does not affect the radiation
scattered by the particle. Furthermore, it must be very stable so
that the particle does not move with respect to the incident wave-
front. There are a variety of methods that have been developed to
support a single particle, such as electric field support of a
charged particle (Egan, 1954), acoustic field support (Fahlen and
Bryant, 1966), support by means of radiation pressure (Ashkin, 1970),
and the use of spider threads (Dessens, 1947; Saunders, 1970). In
this paper the spider thread method is used to support, in an electric
field, a droplet formed by the condensation of water vapor on a
crystal of sodium chloride, and measurements of the backscattered
signal are made as a function of the electric field. Since it is

known that the shape of the droplet is affected by the field, these
measurements should shed some light on the backscattering properties
of droplets that have been slightly deformed from a sphere. Also,
recent theoretical work (Chýlek *et al.*, 1976; Acquista, 1978) has
applied a modified form of Mie theory to equivalent spherical particles in an attempt to deduce the scattering properties of nonspherical particles, the major assumption in this approach being that
irregular particles do not support surface waves while spheres do.
Photographs are presented showing backscattered radiation from a
sodium chloride crystal as the condensation of water vapor progresses.
These photographs suggest that an irregular particle, surrounded by
a thin layer of water, may support surface waves.

A schematic diagram of the electric field experiment is shown
in Figure 1. The spider threads were obtained from a small spider
found in the laboratory and were wound on a tensioning frame
(Saunders, 1967). A sodium chloride solution was sprayed on the
threads after which the threads and crystals (the water having
evaporated) were examined under a microscope. All of the threads
were removed from the frame except that thread that was to be used
for supporting the droplet. This thread was just at the limit of
detectability when examined by backscattered light from a helium-
neon laser, the small crystals serving to delineate the thread. The
experiment requires that only one droplet be supported on the thread.

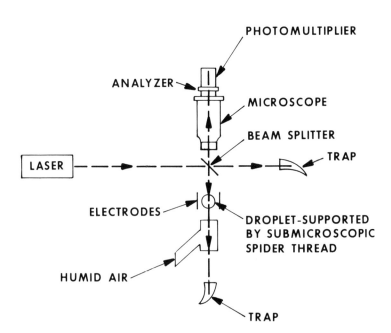

FIG. 1.   Schematic of electric field scattering experiment.

Consequently, the crystals were turned into droplets by directing a flow of water vapor at the crystals, and by means of a micromanipulator, a probe was caused to touch each unwanted droplet. As the probe was withdrawn the droplet separated from the thread. In this manner droplets were removed from the thread except for one droplet located near the center of the thread.

A two-channel eyepiece is used in the microscope in Figure 1. One channel directs the backscattered light to a photomultiplier, the second channel directs a fraction of this light to a television camera so that the surface wave component of the backscattered light and the geometrical optics component (the light reflected from both poles of the sphere) can be monitored. If the dimensions of the droplet are changing, the geometrical optics component will vary in intensity due to interference effects (Fahlen and Bryant, 1968). The photomultiplier signal was recorded when the dimensions of the droplet were essentially constant.

Figure 2 shows the total backscattered signal as a function of the electric field. The vertical bars represent the maximum and minimum signals. The droplet in a zero electric field appeared to be spherical with a diameter of 42 μm. No eccentricity could be detected by means of photography, and no photographs were taken as the field was varied since O'Konski and Thatcher (1953) have indicated that, for droplets of the size used in this experiment and for the magnitude of the field used, no photographic measurement would reveal an asymmetry. The increase in signal to a maximum, followed by a decrease, was repeatable. I conjecture that this behavior is due to a gravitational distortion and I assume that the maximum signal occurs for a spherical droplet. Initially, with zero electric field, the droplet is elongated in the vertical direction due to gravitational distortion. As the field is increased, we know that the droplet becomes elongated in the horizontal direction. Consequently, there will be a value of field at which the initial gravitational distortion is overcome. For this droplet, this value of electric field is about 200 volts/cm. Therefore, using the theory of O'Konski and Thatcher that gives the eccentricity as a function of the radius of a spherical droplet, the electric field, the surface tension and the dielectric constant of the medium surrounding the droplet, I deduce the values of eccentricity given in Figure 2. To do this, I also assumed that the theory of deformation holds for droplets that are very nearly spherical. From Figure 2, the implication is that extremely small changes in droplet eccentricity can cause large variations in the backscattered signal.

Figure 3 is a sequence of photographs showing helium-neon light backscattered from a crystal of sodium chloride as water vapor condenses on the crystal. Figure 3a shows the crystal before condensation takes place and the sequence from Figure 3b through 3e shows the appearance of the crystal as the amount of condensed water

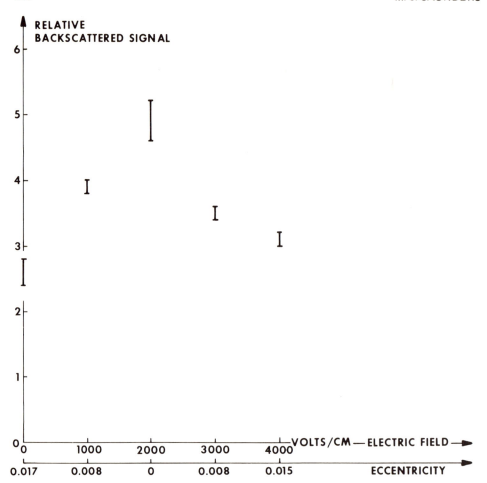

FIG. 2. Backscattered signal vs. electric field strength droplet diameter 42 μm.

increases. The dimensions of the cross section of the original crystal are about 58 μm by 60 μm. Figure 3f is a photograph of the equilibrium droplet (diameter = 119 μm) formed from the crystal. The incident light is polarized parallel to the scattering plane and an analyzer in the eyepiece is crossed with respect to the incident polarization direction to remove virtually all of the geometrical optics component. Photograph 3d seems to indicate the presence of a surface wave coming from an irregularly shaped particle surrounded by a thin layer of water. Consequently, although irregularly shaped particles probably do not support surface waves, irregularly shaped particles surrounded by a thin water sheath may support these waves. Therefore, for these particles, scattering calculations

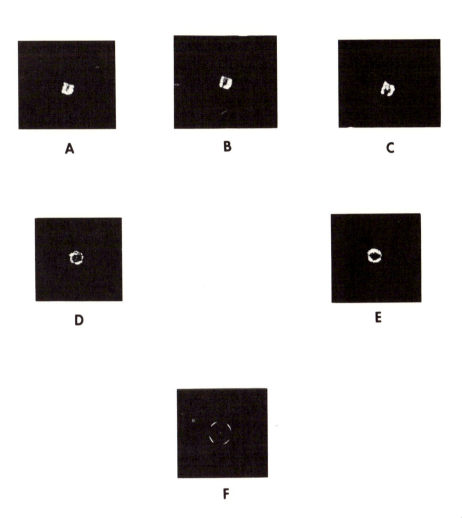

FIG. 3. Photographs of sodium chloride crystal in backscattered helium-neon light as water vapor condenses on crystal. Water vapor condensation increases a → e. Inset 3f shows equilibrium droplet.

in which the surface wave contribution from the Mie theory is removed may lead to erroneous results.

REFERENCES

Acquista, C., 1978, *Appl. Opt.* $\underline{17}$, 3851.
Ashkin, A., 1970, *Phys. Rev. Lett.* $\underline{24}$, 156.
Chýlek, P., Grams, G. W., and Pinnick, R. G., 1976, *Science* $\underline{193}$, 480.
Dessens, H., 1947, *Météorologie 1947*, (October – December), 321.
Egan, J. J., 1954, thesis, Indiana University.
Fahlen, T. S. and Bryant, H. C., 1966, *J. Opt. Soc. Am.* $\underline{56}$, 1635.
Fahlen, T. S. and Bryant, H. C., 1968, *J. Opt. Soc. Am.* $\underline{58}$, 304.
O'Konski, C. T. and Thatcher, Jr., H. C., 1953, *J. Phys. Chem.* $\underline{57}$, 955.
Saunders, M. J., 1967, *Science* $\underline{155}$, 1124.
Saunders, M. J., 1970, *J. Opt. Soc. Am.* $\underline{60}$, 1359.

IN SITU LIGHT SCATTERING TECHNIQUES FOR DETERMINING AEROSOL

SIZE DISTRIBUTIONS AND OPTICAL CONSTANTS

Gerald W. Grams

Georgia Institute of Technology, School of Geophysical Sciences

Atlanta, Georgia   30332

ABSTRACT

Our Atmospheric Optics Group at Georgia Tech has initiated a
research program to improve existing instrumentation and data analysis techniques for obtaining *in situ* data on size distributions and
optical properties of airborne particles. This work is based on the
use of laser polar nephelometers to measure scattering phase functions
for aerosols of known size, shape, and composition to provide input
data for testing the validity of a variety of techniques for calculating scattering by non-spherical particles.

1. INTRODUCTION

Needs exist for basic research in support of the development of
new *in situ* and remote-sensing techniques for characterizing the
optical constants for natural and artificial aerosol particles. We
have initiated a three-year program to use laser polar nephelometers
for observations of angular scattering functions for laboratory-generated aerosol particles of known refractive index. The shapes and
size distributions of the particles will be documented by a number
of techniques including the use of single-particle optical counters
and the analysis of scanning electron microscope photographs of particles collected onto filters. This data will then be used for testing
a variety of techniques for deriving size distributions and optical
properties of aerosol particles.

This program will involve several parallel efforts. For example,
theoretical predictions and experimental observations of light-
scattering functions for the laboratory-generated particles will be

intercompared to establish the effects of particle size, shape, and refractive index on the angular scattering patterns. As these results become available, they will be used as input data for developing and testing inversion algorithms for determining aerosol optical parameters from the light-scattering data. At the same time, our laboratory laser nephelometers will be modified and improved to provide field-worthy instrumentation for *in situ* measurements of the optical properties of airborne particulates such as dusts, fogs, and smokes.

2. LASER NEPHELOMETER SYSTEMS

Our Atmospheric Optics Laboratory includes several unique instruments for carrying out research on light scattering by aerosol particles. Our laser polar nephelometer (Grams *et al.*, 1976) is a compact device capable of being operated in the laboratory, in an instrumented trailer for field measurement programs, or in pressurized aircraft for airborne surveys of aerosol particle parameters. The device uses a collimated laser beam as the light source. An optical system defines a narrow field of view (0.5 degree half-angle). A photomultiplier photon-counting system is used to measure the photomultiplier pulse rate with the light beam both on and off; the difference in the measured pulse rates is directly proportional to the intensity of the light scattered from the volume common to the intersection of the laser beam and the detector field of view. Measurements can be made at scattering angles of 10 to 170 degrees from the direction of propagation of the light beam. Intermediate angles between these extremes are obtained by selecting the angular increments desired. Pulses provided by digital circuits control a stepping motor which sequentially rotates the detector by preselected angular increments (usually 5 degrees, although any multiple of 0.1 degree may be used). We are also developing an improved version of this device for installation on NASA's U-2 aircraft. This instrument extends the angular range of the measurements to cover scattering angles from 5 to 175 degrees. When completed, the U-2 nephelometer will also be available for use in our research program. Both of these polar nephelometers can be operated manually or under the control of our Data General NOVA 1200 minicomputer data management system. The minicomputer data system has been used with our laser nephelometer in almost a dozen different ground-based or airborne field measurement programs. The U-2 nephelometer will incorporate a microprocessor for instrument control and data recording.

We have also constructed a "breadboard" polar nephelometer system which uses a large optical table for mounting various lasers and detectors in different optical configurations for testing new approaches for measuring the angular scattering patterns. The present configuration of the breadboard optical system includes a high-power argon-ion laser for pumping a continuous-wave dye laser system. The dye-laser output beam can then, for example, be directed through our airborne nephelometer for studies of the spectral variation of angular scattering cross sections of aerosols.

We also have instrumentation and data analysis techniques for determining the imaginary refractive index of aerosol samples for visible radiation by measuring the total diffuse reflectance of a sample containing the aerosol and applying Kubelka-Munk theory to the reflectance data (see, *e.g.*, Patterson *et al.*, 1977). Reflection measurements for Kubelka-Munk analyses are made with a Cary-14 spectrophotometer equipped with a 15-cm integrating sphere coated with barium sulfate paint; the equipment and measurement procedures are similar to those used by Lindberg and Laude (1974). A capability for determining the imaginary refractive index of collected particles using standard laboratory procedures will enable us to carry out intercomparisons with *in situ* techniques.

As a means of documenting particle size and shape information for aerosol particles of known composition generated under controlled laboratory conditions, we will collect the artificially generated particles onto Nuclepore polycarbonate membranes for analysis on the scanning electron microscope (SEM) at Georgia Tech's Engineering Experiment Station. Examples of the use of SEM photographs for determining particle size and shape are given in Figure 1.

3. RESEARCH PLANS

Our laboratory program will place a strong emphasis on light scattering by randomly oriented irregular particles. This effort will be an extension of the work described by Chýlek *et al.* (1976). In that study, a polar nephelometer was used for a series of laboratory observations of light scattering by laboratory-generated particles to establish the value of a parameter that would give best agreement between measured scattering coefficients and those calculated using the modified Mie technique described by Chýlek (1975). As shown in Figure 1, Chýlek's *ad hoc* procedure provided good agreement over the range of particle size and refractive index that was used.

We will extend the above study to a wider range of refractive indexes and particle sizes; in particular, we will study laboratory generated particles with Mie size parameters in the range from approximately 1 to 50, and complex refractive indexes having an imaginary part in the range from about 0.001 to at least 0.1. We also plan to use our laboratory data to evaluate other techniques for calculating scattering by non-spherical particles such as the *ad hoc* procedure suggested by Pollack and Cuzzi (1978) and exact solutions for non-spherical particle shapes such as those described by Barber and Yeh (1975), Ward (1974), or Weil and Chu (1976).

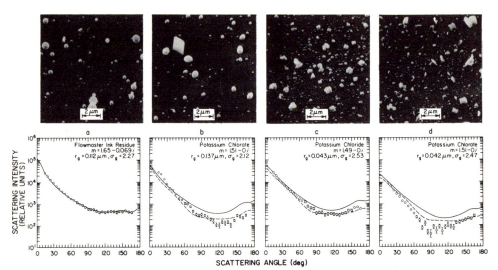

FIG. 1.   Comparison of measured angular scattering patterns at
λ = 0.6328 μm (circles with error bars) with those calculated by Mie
theory (solid lines) and a modified form of the Mie theory (Chýlek,
1975).   Particle composition and complex refractive index are identified on each scattering diagram; each diagram also contains the
geometric mean radius r_g and geometric standard deviation σ_g for
log-normal size distributions obtained by determining the radii of
spheres of equal cross sections for several thousand particles from
scanning electron microscope photographs.   Examples of such photographs are presented above the scattering diagrams to illustrate
their respective particle shapes (from Chýlek *et al.*, 1976).

REFERENCES

Barber, P. W. and Yeh, C., 1975, *Appl. Optics*, *14*, 2864.
Chýlek, P., 1975, *J. Opt. Soc. Am.*, *66*, 285.
Chýlek, P., Grams, G. W., and Pinnick, R. G., 1976, *Science*, *193*, 480.
Grams, G. W., Dascher, A. J., and Wyman, C. M., 1975, *Opt. Eng.*, *14*,
    85.
Lindberg, J. D. and Laude, L. S., 1974, *Appl. Optics*, *13*, 1923.
Patterson, E. M., Gillette, D. A., and Stockton, B. H., 1977, *J.
    Geophys. Res.*, *82*, 3153.
Pollack, J. B. and Cuzzi, J. N., 1978, *in* "Scattering by Non-Spherical
    Particles of Size Comparable to a Wavelength: A New Semi-Empirical
    Theory," Preprint volume, Third Conf. on Atmospheric Radiation,
    28-30 June 1978, Davis, California; published by the American
    Meteorological Society, Boston, Mass.
Ward, G., 1974, *in* "Electromagnetic Scattering from an Irregular Di-
    electric Particle," Proc. IEEE, Southeastcon 1974, Region 3 Confer-
    ence, 29 April - 1 May 1974, Orlando, Florida, p. 160.
Weil, H. and Chu, C. M., 1976, *Appl. Optics*, *15*, 1832.

PROBLEMS IN CALIBRATING A POLAR NEPHELOMETER

Alfred Holland

NASA, Wallops Flight Center

Wallops Island, Virginia 23337

ABSTRACT

A calibration method based on Pritchard's original work is outlined. Results for an existing nephelometer system are discussed.

In studying the scattering behavior of small particles the polar
nephelometer is a powerful tool. Basically a nephelometer consists
of a source telescope that is fixed in relation to a circular mounting
base, or to a central hub. A scanning arm carrying a receiver telescope is pivoted at the center of the mounting base or at the hub.
Both the source and the receiver telescopes are fitted with spectral
filters, polarizers and analyzers. A light trap mounted opposite the
receiver telescope provides a uniform black background for the receiver. The intersection of the source beam and receiver field of
view define the common volume or scattering volume of the instrument,
while the scattering plane is defined by the mounting base or by the
source and receiver arms. Figure 1 shows the essential features of
a typical nephelometer.

Nephelometers have been in use since 1945, and the first description of a nephelometer was furnished by Waldram (1945) who also
provided the name, derived from the Greek *nephelo* or cloud. The
nephelometer was further developed by Mueller (1947), who also developed the Mueller Phenomenological Algebra (the Mueller or scattering matrix). Further work on the interpretation of nephelometer
measurements was carried out by Parke (1949) and by Jones (1947).
Pritchard and Elliott (1960) analyzed the nephelometer and showed
how to optimize its performance. They fabricated a portable polar

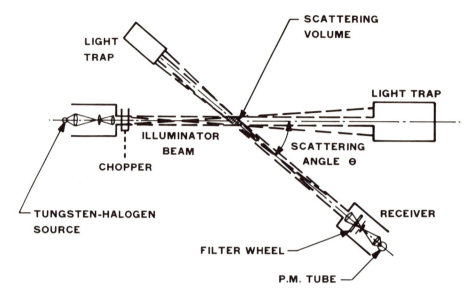

FIG. 1.   A schematic of a polar nephelometer.

nephelometer and used it to study the scattering properties of
natural fogs. However, their most significant contribution to the
field was the calibration method that they developed.

Their analysis of nephelometers showed that the best performance
(balance between sensitivity and angular resolution) was obtained
with a focused system in which both the source lamp filament and the
receiver photocathode were imaged at the scattering volume. The best
compromise is achieved when the field and objective apertures are
higher than they are wide and when each beam is roughly the same size
at the sample space as at the objective lens of the telescope. The
rectangular nature of the source and receiver beams also aids in
determining the size of the sample volume when calibrating the nephelometer.

The calibration procedure developed by Pritchard and Elliott
essentially compared the intensity of the light scattered from the
particles in the common volume with the intensity reflected from a
calibration standard. The standard, made of thin plastic material,
is placed in the common volume perpendicular to the scattering plane
of the instrument (Fig. 2). The screen is moved slowly in a direction perpendicular to the screen surface, along the major diagonal
of the common volume, so that each pair of rays common to the source
and receiver beams is intercepted during the screen travel. The
intensity of light reflected by the screen is recorded during its
movement, and the integral of the signal over the distance traveled

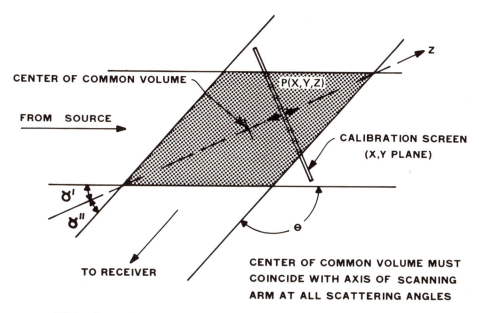

FIG. 2.   A schematic of the calibration procedure.

is calculated (Fig. 3). That integral is compared to the intensity
scattered by the particles in the common volume during scattering
measurements. The reflection and transmission matrices of the calibration screen are also measured and compared to an absolute standard - a glass plate heavily covered with magnesium oxide particles,
whose reflection matrix was determined by Pritchard and Elliott for
angles near the normal in the visible spectrum.

     Figure 3 also provides additional information. If the peak
signal always occurs at the center of the common volume (the zero
point on the calibration carriage), then the source and receiver
beams are in correct alignment. The length of the major diagonal
of the common volume is represented by the length of the ordinate
for which the signal is not zero.

     A proof of the calibration method is straightforward and follows
Pritchard and Elliott's exposition with a few minor changes. In
Figure 2 the source beam is shown proceeding from left to right,
with the calibration screen positioned so that the normal to the
screen surface lies along the major diagonal of the common volume.
An area dxdy about point $P(x,y)$ on the illuminated portion of the
screen receives a flux of $E(x,y,z) \cos \gamma'$ dxdy (watts cm^{-2}) normal
to its surface. The notation $E(x,y,z)$ allows for variations in the
source beam over its aperture and variations with distance along the
beam. If the screen is a Lambert reflector with a normal reflectance

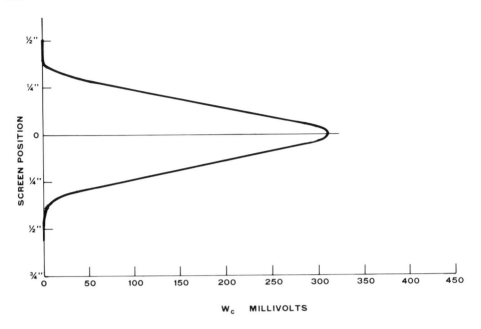

FIG. 3.   A typical calibration signal.

coefficient R_N, then the intensity reflected into the receiver is

$$dI = \frac{R_N}{\pi} E(x,y,z)dA \cos \gamma' \cos \gamma'' \left(\frac{watts}{steradian}\right), \text{ where } dA = dxdy.$$

The flux into the receiver will be linearly proportional to the reflected intensity:

$$dF = C_1 \frac{R_N}{\pi} E(x,y,z)dA \cos \gamma' \cos \gamma'' \text{ (watts)},$$

where the constant depends on the receiver aperture and the distance from the screen to the aperture and is a function of the coordinates (x,y,z). Similarly the signal recorded by the electronic system will be linear with the input flux:

$$dW_c = C_1 C_2 \frac{R_N}{\pi} E(x,y,z)dA \cos \gamma' \cos \gamma'' \text{ (millivolts)}.$$

Integrating over the illuminated portion of the screen we have

$$W_c = \frac{R_N}{\pi} \cos \gamma' \cos \gamma'' \int_A \int C_1 C_2 E(x,y,z)dA \quad (\text{millivolts}) ,$$

and integrating over the axis of screen travel, z, we have the calibration equation:

$$\int_Z W_c \, dZ = \frac{R_N}{\pi} \cos \gamma' \cos \gamma'' \int \int \int_V C_1 C_2 E(x,y,z)dV \quad (\text{millivolts-cm})$$

where dAdZ = dV. Since the coefficients C_1 and C_2 are allowed func-
tional dependence on the spatial variables, we can account for variations in photocathode response and variations in distance from screen
to receiver aperture.

Now when a scattering measurement is made, the calibration
screen is removed, and a scattering medium fills the common volume.
The experimental conditions are identical to the calibration conditions so that an element of volume dV about the point P(x,y,z)
scatters an intensity

$$dI (\theta) = \beta_s(\theta) E(x,y,z)dV \quad (\text{watts ster}^{-1})$$

into the receiver telescope set at scattering angle θ. Since the
same optical system is used, the flux into the receiver is

$$dF_s(\theta) = C_1 \, dI_s(\theta) \quad (\text{watts}) ,$$

while the response of the system is

$$dW_s(\theta) = C_1 C_2 \, dI_s(\theta) = C_1 C_2 \beta_s(\theta) E(x,y,z)dV \quad (\text{millivolts}) .$$

The parameter $\beta_s(\theta)$ is the volume scattering coefficient of the
scattering medium in cm^2/cm^3. Integration over the common volume
gives

$$W_s(\theta) = \beta_s(\theta) \int \int \int C_1 C_2 E(x,y,z)dV \quad (\text{millivolts}),$$

where it is assumed that $\beta_s(\theta)$ is constant over the small range of scattering angles accepted by the receiver at each scattering angle θ. The angular resolution of our present nephelometer system is $< 1°$ in scattering angle.

Solving for the scattering coefficient we have

$$\beta_s(\theta) = W_s(\theta) \Big/ \int \int \int_V C_1 \, C_2 \, E(x,y,z)dV \quad (cm^{-1}) \, ,$$

and substituting for the volume integral from the calibration equation, we finally have

$$\beta_s(\theta) = W_s(\theta) \, \frac{R_N \, \cos \gamma' \, \cos \gamma''}{\pi \int_Z W_c(\theta) \, dZ} = W_s(\theta) \, K(\theta) \quad (cm^{-1}) \, .$$

The parameter $K(\theta)$ is the calibration constant for scattering angle θ. Notice that if $E(x,y,z) = E$ is constant and C_1 and C_2 do not vary then the volume integral reduces to

$$\int \int \int_V C_1 \, C_2 \, E(x,y,z)dV = C_1 \, C_2 \, E \, V \, ,$$

and for rectangular beams the volume is a parallelepiped whose volume is inversely proportional to $\sin \theta$, which often holds for real systems. Figure 4 shows a typical calibration curve for our polar nephelometer system. In practice it is necessary to obtain values of the calibration constant for each polarization state of the source and receiver, for each spectral passband, and for a sufficient number of scattering angles to define the functions. But if we denote the calibration coefficient for the polarized case as $K(P,A,\theta)$ where P denotes the source polarizer and A the receiver analyzer and U denotes the unpolarized case, simple calculations will show that for the same spectral passband

$$K(P,A,\theta) = \frac{1}{a_{11}p_{11}} \, K(U,U,\theta) \, .$$

Here a_{11}, p_{11} denote the multiplier of the normalized analyzer and polarizer matrices. For example,

$$A_{ij} = a_{11} \begin{vmatrix} 1.0 & a'_{12} & a'_{13} & a'_{14} \\ a'_{21} & a'_{22} & a'_{23} & a'_{24} \\ a'_{31} & a'_{32} & a'_{33} & a'_{34} \\ a'_{41} & a'_{42} & a'_{43} & a'_{44} \end{vmatrix}.$$

The values a_{11}, p_{11} can be obtained at each wavelength by using the MgO coated plate in reflection, since the reflection matrix for MgO powders is known (Pritchard and Elliott, 1960).

In practice a secondary standard is used in calibration for both reflection ($180° > \theta > 90°$) and transmission ($0° < \theta < 90°$), and its properties are determined by comparison to a magnesium oxide coated standard in reflection. The fact that the calibration constant must be identical at 90° whether determined in reflection or transmission allows the reflection calibration to be continued to transmission angles.

I have dwelt on the aspects of calibration at some length because of its importance in nephelometry. For example, failure to take into account the change in the common volume with scattering

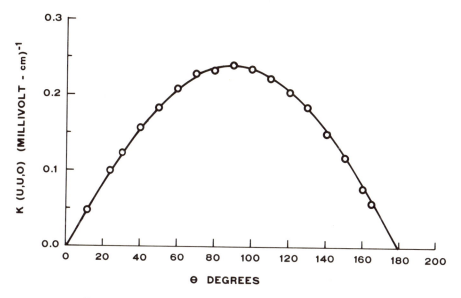

FIG. 4.   A completed calibration curve.

angle can lead some investigators to find a glory in the backscattering pattern of non-spherical particles. Such false conclusions could
not be reached if the data were corrected by the instrument calibration constant which is proportional to the sine of the scattering
angle.

REFERENCES

Jones, R. C., 1947, *J.O.S.A.*, <u>37</u>, 107.
Mueller, H., 1947, *in* "Theory of Polarimetric Investigations of
  Light Scattering," Parts I and II, Contract W-18-035-CWS-1304,
  D.I.C. 2-6467, M.I.T.
Parke, N. G., 1949, *in* "Statistical Optics: Mueller Phenomenological
  Algebra," Research Laboratory of Electronics, (M.I.T) Rpt. #119.
Pritchard, B. S. and Elliott, W. G., 1960, *J.O.S.A.*, <u>50</u>, 191.
Waldram, J. W., 1945, *Trans. Illum. Eng. Soc. (London)*, <u>10</u>, 147.

EXTINCTION SIGNATURES OF NON-SPHERICAL/NON-ISOTROPIC PARTICLES

R. T. Wang

State Univ. of N. Y. at Albany, Space Astronomy Laboratory

Albany, New York 12203

ABSTRACT

Microwave analog measurements of $\theta = 0°$ scattering by particles of various shape and refractive indexes are presented. The scattering targets are non-isotropic spheres, aggregates of 2^n (n=1,2,3) identical spheres, and stacked 7-cylinder rough particles. Both the orientation and the physical properties of the target produce signatures in the extinction and polarization curves. Theoretical explanations are presented wherever possible.

1. INTRODUCTION

A small particle obscures (or dims) the incident light through a subtle interference phenomenon between its forward-scattered ($\theta = 0°$) wave and the incident wave. This interpretation of extinction leads to the well-known Optical Theorem (Feenberg, 1932; van de Hulst, 1946, 1949, 1957; Montroll and Greenberg, 1954),

$$C_{EXT} = \frac{4\pi}{k^2} |S(0)| \sin\phi(0) = \frac{4\pi}{k^2} \text{Re}\{S(0)\} \quad , \tag{1}$$

which relates the extinction cross section C_{EXT} to the real part of the dimensionless complex amplitude $S(\theta)$ along the incident direction ($\theta = 0°$) and polarization, or equivalently, to the absolute magnitude $|S(0)|$ and the phase shift $\phi(0)$ of the forward scattered wave.

Each P,Q plot presented in this paper is a cartesian representation of S(0) as a function of particle orientation angle χ; *i.e.*, in the complex plane the dimensionless P and Q components are:

255

$$P = \frac{4\pi}{k^2 G} \, \text{Im}\{S(0)\} \quad ; \quad Q = \frac{4\pi}{k^2 G} \, \text{Re}\{S(0)\} \tag{2}$$

where G is the appropriate geometrical cross section of the particle
(or ensemble of particles) and Q is the so called "extinction
efficiency". As the orientation of the particle changes, a curve is
generated in this plane. The vector from the origin to a specific
point (particle orientation) on the curve represents $S(0)$ while the
angle between this vector and the P axis corresponds to $\phi(0)$. The
absolute magnitude $|S(0)|$ is calibrated against that of a standard
sphere of known $|S(0)|$ in the same P,Q plot. The projection of $S(0)$
into the Q axis gives the extinction efficiency. For a particle of
rotational symmetry, the totality of the extinction information is
obtained by rotating the symmetry axis through 90° in two mutually
orthogonal planes, the k-E and the k-H plane of the incident wave.
For example, if the particle axis makes an arbitrary angle χ from
the incident direction $\vec{k}$, $S(0)$ is the simple linear composition of
those values of $S(0)$ obtained when the symmetry axis is in these two
planes (k-E and k-H), tilted by χ from $\vec{k}$. Furthermore, if one keeps
χ fixed and sweeps the axis on a cone around $\vec{k}$, the tip of the $S(0)$
vector draws a straight line in the P,Q plot. These remarkable.
scattering properties at $\theta=0°$, which are the result of mathematical
symmetry and independent of target material, considerably reduce the
number of required measurements. It has been our experience that
this property of simple linear composition is nearly true even for
particles of less rotational symmetry like helixes (Wang, 1970).
More details on information necessary to interpret the experimental
plots are contained in the rather lengthy figure captions and earlier
publications (Lind *et al.*, 1965; Lind, 1966; Greenberg *et al.*, 1967;
Wang, 1968; Wang *et al.*, 1977; Wang and Greenberg, 1978).

Extinction is perhaps the least sensitive scattering measurement
for distinguishing among anisotropic particles, especially when they
are randomly oriented. Nevertheless, there exist conspicuous,
systematic differences in the $\theta=0°$ scattering which seem to provide
a (unique?) particle signature. This is particularly true as a
function of particle orientation as seen in the discussions to
follow. Remember the following rules for interpreting each P,Q
plot. *A vector drawn from the coordinate origin to each χ position
along the curve yields the complex value of $S(0)$ at χ. The phase
shift $\phi(0)$ is given by the angle between this vector and the P axis.
The extinction efficiency (C_{EXT}/G) is found at χ by projecting this
vector on the calibrated Q axis. The length of the vector represents
the absolute value $|S(0)|$, and its numerical value is obtained by
comparing this length with that of the "standard" or calibration
vector (obtained from a sphere) provided in each plot. The proper
numerical value of $|S(0)|$ for the calibration sphere is given in the
figure caption. The geometrical cross section G of an aggregate of
spheres is taken to be the sum of those of the component spheres.*

2. EXPERIMENTAL RESULTS & COMPARISON WITH THEORETICAL PREDICTIONS

2.1 Non-isotropic Spheres

A medium composed of thin alternating layers of different electromagnetic properties has anisotropic refractive indexes with respect to the polarization of the propagating wave (Born and Wolf, 1965; Rytov, 1955). By inserting an absorbing (conducting) film into each interface of neighboring dielectric (expanded polystyrene) layers, spheres of various anisotropies have been constructed for extinction studies (Wang, 1968; Wang and Greenberg, 1976). P,Q plots resulting from two such spheres are shown in Figures 1A and 1B. These two spheres differ from each other only in their inserted films. One has the absorbing film called "teledelto" while the other has aluminum foil. In both cases the axis of symmetry passes through the center, perpendicularly to the layer planes. These spheres are characterized by three indexes of refraction, each measured along one of three principal directions. Based on reasons similar to those used in the Eikonal approximations (van de Hulst, 1957; Greenberg, 1960; Wang and Greenberg, 1976), one can construct an effective index m_χ for each particle orientation angle. By applying the rigorous Mie solution for isotropic spheres with this adjusted m_χ , one obtains a fairly close theoretical prediction for the χ-dependence of $S(0)$.

For the anisotropic sphere represented in Figure 1A, m_χ was computed according to the Fresnel formula of crystal optics (Born and Wolf, 1965):

$$\frac{1}{m_\chi{}^2} = \frac{\cos^2\chi}{m_k{}^2} + \frac{\sin^2\chi}{m_E{}^2} \qquad \text{(particle-axis in the k-E plane)}$$

$$\tag{3}$$

$$\frac{1}{m_\chi{}^2} = \frac{\cos^2\chi}{m_k{}^2} + \frac{\sin^2\chi}{m_H{}^2} \qquad \text{(particle-axis in the k-H plane)}$$

The ensuing Eikonal-Mie prediction is shown in the P,Q plot as a dotted curve. It agrees closely with the experimental $S(0)$ vs χ curve, expecially in the E orientation where the particle axis is parallel to the incident $\vec{E}$ vector and where the most precise re-fractive-index measurement of the medium is possible. The same technique has been used in Figure 2 to display the degree of polar-ization by extinction as a function of χ for 3 spheres of the same type of anisotropy. A comparison with non-spherical particles such as spheroids in the same size range (Wang and Greenberg, 1978) demonstrates that the anisotropic sphere is a more efficient polar-izer in its orientation dependence.

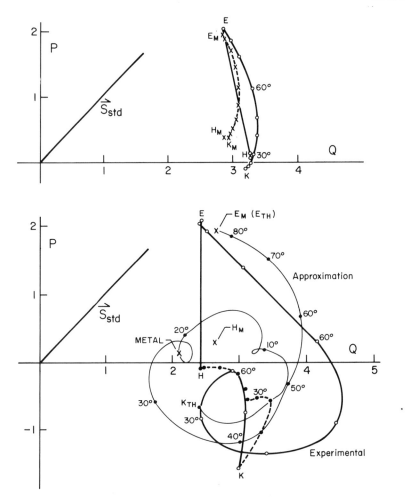

FIG. 1A(top) and 1B(bottom).   Two P,Q plots showing the target
orientation dependence (χ given in deg) of S(0) for two layered-
spheres.   Fig. 1A refers to the teledelto-layered sphere (x=4.86);
the bottom figure corresponds to a similar (x=4.83) sphere but
layered with aluminum foil.   The refractive indexes as measured along
the three principal target media directions are (Fig. 1A)
$m_k = 1.39-i0.14$, $m_E = 1.28-i0.005$, $m_H = 1.38-i0.15$; (Fig. 1B)
m_k = metallic, $m_E = 1.27-i0.009$, $m_H = 1.62-i5.44$.   At the orientation
marked k, the particle symmetry axis is parallel to the incident
direction, $\vec{k}$.   It is then continuously swept through 90° in the k-E
and k-H planes of the incident wave to display two (thick) experimental curves.   The Eikonal-Mie theory predictions are shown either
by a dotted or by a thin curve.   In particular, k_M, $E_M(E_{TH})$, H_M and
Metal are Mie theory results using m_k, m_E, m_H and $m=\infty$, respectively.
For the standard sphere, $|S(0)| = 13.57$; see the last paragraph of
section 1.

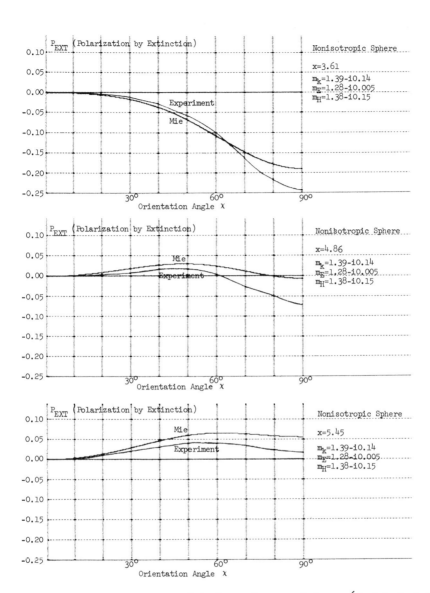

FIG. 2.   Target-orientation (χ) dependence of $P_{EXT} = \left(Q_E(\chi) - Q_H(\chi) \right) /$
$\left(Q_E(\chi) + Q_H(\chi) \right)$, the polarization by extinction, for three spheres of
non-isotropic refractive indexes. $Q_E(\chi)$ and $Q_H(\chi)$ are extinction
efficiencies when the symmetry axis is in the k-E and in the k-H
planes, respectively, tilted by an angle χ from the incident direction
$\hat{k}$.   The results of Eikonal-Mie theory prediction are also shown.

The Eikonal-Mie approximation is applicable only for the k-E rotation in a sphere of enhanced anisotropy as in Figure 1B. The first empirical finding (Wang, 1968) showed the amazing fact that at E orientation the incident wave almost did not see the aluminum foils that separated the dielectric layers! The penetrating wavelets are guided along each dielectric layer independently of each other. As the particle-axis is tilted by $\pi/2-\chi$ from the $\vec{E}$ direction toward $\vec{k}$, each ray of length L in the particle has to traverse an additional distance of $L(1-\sin\chi)$ in free space to cover the same distance in the $\vec{k}$ direction as that in the E orientation. This consideration gives the effective m_χ as

$$m_\chi = m_E + 1 - \sin\chi. \tag{4}$$

Application of the Mie theory prediction with this m_χ is shown as a thin curve, E_{TH} to k_{TH} in Figure 1B. The agreement with observation is not as good as in Figure 1A, but qualitatively predicts the changes in magnitude and phase of $S(0)$ as the particle rotates. For rotation in the k-H plane, the layer structure exhibits a waveguide cutoff phenomenon, and the Eikonal picture loses its meaning. No progress has been made in the explanation of this phenomenon.

2.2 2^n Identical Spheres; n=1, 2, 3

A number of polystyrene spheres have been made from three different sized metallic molds. For each size, spheres of nearly the same measured $|S(0)|$ and $\phi(0)$ were selected for this dependent scattering study. The refractive index of these particles resembles that of water or ice in the optical region. For each size group the phase-shift parameter $\rho = 2x(m'-1)$ and extinction efficiency of a single sphere are, respectively, $\rho_1 = 2.278$, $Q_1 = 2.278$; $\rho_2 = 2.746$, $Q_2 = 2.985$ and $\rho_3 = 3.396$, $Q_3 = 3.753$. 2, 4 or 8 of these identical spheres were assembled on nylon strings to form an array of simple geometry, and the array was rotated in the incident-wave k-H plane to display a P,Q plot.

A quick glance at Figures 3A-3I for a 2-sphere array at various center-to-center separations (s) reveals several interesting features: a) The tip of the $S(0)$ vector draws a clockwise spiral as the array axis rotates from the $\vec{k}$ to the $\vec{H}$ direction of the incident wave, with an overall decreasing trend in $\phi(0)$. Prolate spheroids of similar target-parameters have the same trend (Wang and Greenberg, 1978) except that the $S(0)$-tip tends to go counterclockwise. b) In the H orientation, the $S(0)$ vector is nearly equal to the vector sum of those of 2 spheres, *i.e.*, dependent scattering is at a minimum. c) As s increases, the size of the spiral shrinks, perhaps nonmonotonically. This is accompanied by the faster convergence of the

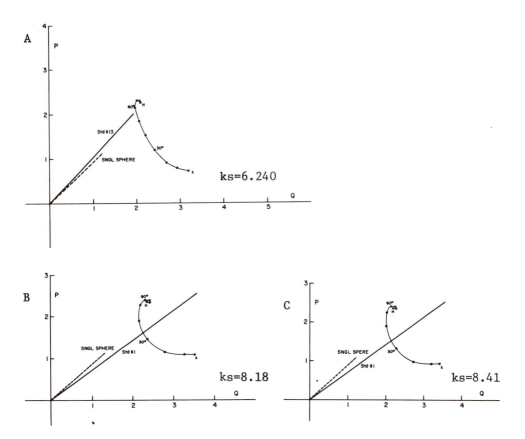

FIGS. 3A-3C.   Experimental P,Q plots of the target-orientation (χ)
dependence of S(0) of an ensemble of two identical spheres
($x_1=x_2=3.120$, $m_1=m_2=1.365$) at various separations $ks=2\pi s/\lambda$.   s is the
center to center distance between two spheres.   At the orientation
position marked k, the symmetry axis of the array is parallel to the
incident direction $\hat{k}$.   This axis is then continuously swept through
90° in the k-H plane of the incident wave to display the experimental
curve.   A dotted straight line marked SNGL SPHERE is the S(0) vector
of an isolated sphere from the ensemble, as measured during the same
experimental run.   The calibration vectors have $|S(0)|=21.38$ (std #1)
and $|S(0)|=13.57$ (std #13); see the last paragraph of section 1.

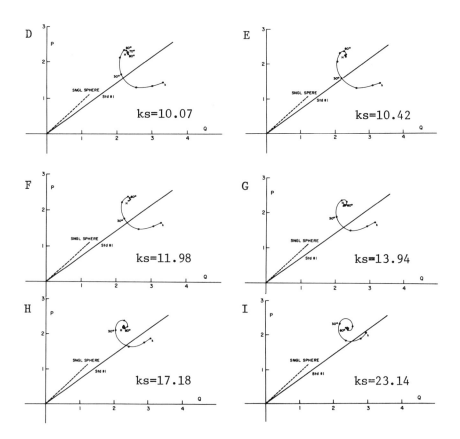

FIGS. 3D–3I.   Same as Figs. 3A–3C.

S(0)-tip toward the H orientation with respect to the χ variation. Figures 3A-3I thus render a vivid picture of how dependent scattering diminishes with increasing s.

A combination of ray optics and Mie theory is proposed here to explain the above subtle χ dependence of S(0). In this approximation the total S(0) from a 2-sphere array (spheres A & B) having a separation s and an orientation χ takes the form:

$$S(0) = S_{1A}(0)+S_{1B}(0) + \frac{S_{1A}(\chi)S_{1B}(\chi)}{iks} e^{-iks(1-\cos\chi)}$$

$$+ \frac{S_{1A}(\pi-\chi)S_{1B}(\pi-\chi)}{iks} e^{-iks(1+\cos\chi)} . \tag{5}$$

$S_{1A}(\chi)$ is the perpendicular scattering amplitude component at scattering angle χ from sphere A, and so on for other expressions, as evaluated from Mie theory, all in van de Hulst's (1957) notations. The 1st and 2nd terms represent the summation of independent scattering by spheres A & B. The 3rd term takes account of a multiple-scatter correction to S(0) due to the scattering by sphere B of radiation from sphere A. In going from sphere A to B the near field is approximated by a wave of plane phase front perpendicular to the ray path but with an inverse s dependence in its amplitude. Similarly, the 4th term is that due to the scattering by sphere A of radiation from sphere B. This approximation is shown in Figures 4A-4C for 3 sets of s, and it can also be shown to give the same expression as derived from a rigorous solution using the relocation technique of vector spherical wave functions (Liang and Lo, 1967; Bruning and Lo, 1971) for large values of s. For s < 8a, the approximation overestimates $|S(0)|$ and underestimates $\phi(0)$ for this particular ensemble, especially near the k orientation.

Figures 5A-5C and Figures 6A-6C show the $\theta=0°$ scattering signatures resulting from 4 and 8 spheres forming contacting square and cubic ensembles, respectively. Despite their simple appearance, no detailed theoretical explanation is presently available. Nevertheless, there exist similarities between ensembles made up of the same size spheres. First, note the comparisons between Figures 5A and 6A, 5B and 6B, and 5C and 6C. The signatures are similar except that $|S(0)|$ for the 8-sphere ensemble is about twice the value as that of the 4-sphere ensemble. This implies no appreciable interference between the two arrays of the 8-sphere ensemble that are parallel to the k-H plane. Second, the experimental Q's of each of the 8-sphere ensembles shown in Figures 6A-6C (multiplied by a factor 2 for equivalent normalization) can be compared with the Q's calculated by Mie theory for a sphere equal in volume to each ensemble. For the 3 sets of 8-sphere ensembles used in Figures 6A-6C, the observed minimum

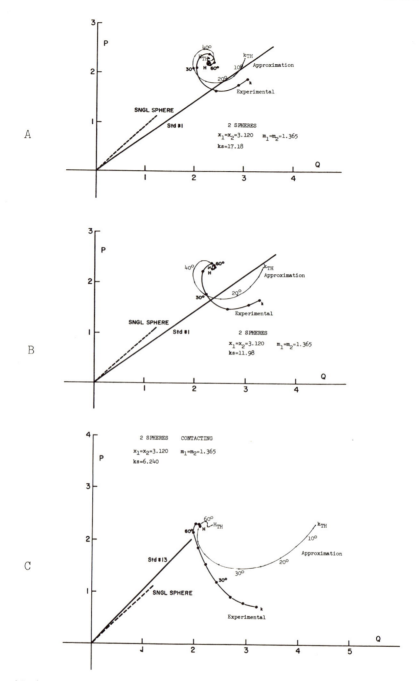

FIGS. 4A-4C.   Same as in Figs. 3A-3I.   Here, the Geometrical Optics
+ Mie Theory prediction is added as a thin curve (k_{TH} to H_{TH}) on each
P,Q plot for 3 sets of mutual separation ks.

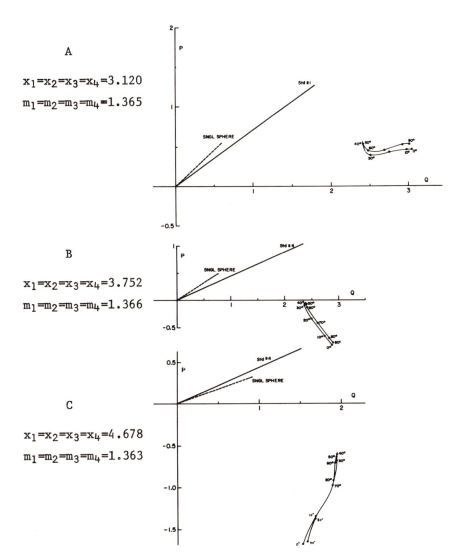

A

$x_1=x_2=x_3=x_4=3.120$

$m_1=m_2=m_3=m_4=1.365$

B

$x_1=x_2=x_3=x_4=3.752$

$m_1=m_2=m_3=m_4=1.366$

C

$x_1=x_2=x_3=x_4=4.678$

$m_1=m_2=m_3=m_4=1.363$

FIGS. 5A–5C.   Three experimental P,Q plots, each of which shows the target-orientation (χ) dependence of S(0) for an ensemble of 4 identical spheres forming a contacting square array. Each plot differs from the others on the size parameter of composite spheres. At $\chi=0°$ (or 90°) one side of the square is parallel to the incident $\vec{k}$ vector, and the square is then continuously rotated in the k–H plane through 90° to display the experimental curve. The dotted straight line marked SINGL SPHERE is the S(0) vector of an isolated sphere from the ensemble, as measured during the same run. $|S(0)|=21.38$ (std #1) and $|S(0)|=36.06$ (std #4); see the last paragraph of section 1.

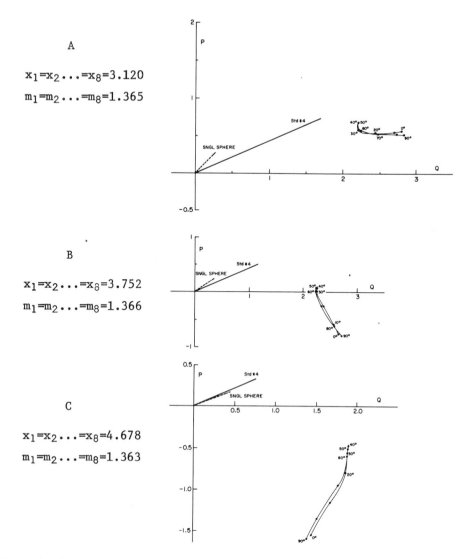

FIGS. 6A–6C.   Three experimental P,Q plots, each of which shows the
target-orientation (χ) dependence of S(0) for an ensemble of 8 identical spheres forming a contacting cubic array. Each plot differs from
others on the size parameter of composite spheres. At $\chi=0°$ (or 90°)
one face of the cube is parallel to the incident $\vec{k}$ vector but is perpendicular to $\vec{H}$. The array is then continuously rotated through 90°
in the k-H plane to display the experimental curve. The dotted
straight line marked SNGL SPHERE is the S(0) vector of an isolated
sphere from the ensemble, as measured during the same run. Std. #4
has $|S(0)|=36.06$; see the last paragraph of section 1.

and maximum Q's, the phase-shift parameters, and Q's for the equivalent spheres are, respectively, Q_{MIN} = 4.41, Q_{MAX} = 5.66, ρ_v = 4.555,
Q_{MIE} = 3.938; Q_{MIN} = 4.49, Q_{MAX} = 5.44, ρ_v = 5.493, Q_{MIE} = 3.131;
Q_{MIN} = 2.77, Q_{MAX} = 3.79, ρ_v = 6.792, Q_{MIE} = 1.920.   This gives an
interesting similarity with the extinction exhibited by the 7-cylinder
rough particles discussed next.

2.3   7-Cylinder Rough Particles

A dielectric rough particle of this type is made by stacking 7
polystyrene (or polystyrene admixed with carbon dust) cylinders of
the same diameter, six of which have a common length/diameter ratio
of 2:1 and symmetrically surround a longer one which has a ratio of
3:1.   By coating such a particle with aluminum foil or by stacking
copper cylinders in the same manner, a totally reflecting particle
without interstices is prepared.   Because the maximum longitudinal
and lateral dimensions of such a particle are the same, it resembles
a roughened sphere.   The size parameter x_v (or the phase-shift
parameter $\rho = 2x_v(m'-1)$) is taken to be that of an equal-volume
smooth sphere.   The ρ or x_v dependencies of extinction efficiency
Q for low-absorbing/reflecting particles at some principal particle-
orientations were published earlier (Greenberg $et~al.$, 1971), and
a more detailed account of some of the P,Q plots has also been
reported (Wang and Greenberg, 1978).   Attention is focused here on
the ρ or x_v dependence of Q_{RAV}, the estimated extinction efficiency
averaged over random particle orientations, and of $\delta Q = Q_{MAX}-Q_{MIN}$,
the range of variation in the observed Q as the particle is rotated.
A few absorbing particles are also included here for a more complete
picture despite some uncertainties in their refractive index which may
amount to an error of ≈0.6 in the ρ scale.   Detailed target parameters and their resulting Q's are listed in Tables 1A and 1B.

The Q's as a function of ρ for 23 such rough particles are
plotted in Figure 7A.   Also shown in the same figure are Mie results
for smooth spheres with m = 1.360-i0.0 and m = 1.360-i0.05.   For each
rough particle, δQ is represented by a vertical bar, and Q_{RAV} is
shown by a horizontal mark.   Some striking contrasts between rough
and smooth particles in this extinction signature are:
  a) Up to $\rho \simeq 2.8$, low-absorbing rough particles extinguish the
  incident light as if they were absorbing equal-volume spheres.
  If $\rho \lesssim 2.2$ this can be qualitatively expected from theoretical
  considerations of the statistical variation of ray-paths passing
  through the particle (Greenberg and Stoeckly, 1970).
  b) δQ tends to increase with ρ and is smaller for an absorbing
  rough particle than for one of the same size with low-absorption.
  c) Near the first major resonance in extinction, $3 \lesssim \rho \lesssim 4.5$,
  the Q_{RAV}'s of rough particles are near those of smooth spheres.
  d) Beyond $\rho \gtrsim 4.5$, Q_{RAV} is consistently larger than Mie results
  for smooth spheres.

TABLE 1A

TARGET PARAMETERS AND EXTINCTION EFFICIENCIES
FOR 7-CYLINDER DIELECTRIC ROUGH PARTICLES

x_v is the size parameter of the equal-volume sphere; m = complex refractive index = $m'-im''$; $\rho = 2x_v(m'-1)$; and Q_{MIN}, Q_{MAX} and Q_{RAV} are the minimum, maximum, and the estimated random average (over particle orientation) of the extinction efficiencies.

Target ID #	Carbon Inclusion	x_v	m'	$-m''$	ρ	Q_{MIN}	Q_{MAX}	Q_{RAV}
022003	0	4.33	1.100	~0	0.87	0.39	0.48	0.47
021005	0	2.83	1.184	~0.003	1.04	0.48	0.56	0.52
023003	0.5%	4.33	1.141	~0.005	1.22	0.85	0.92	0.88
022002	0	6.17	1.101	~0	1.25	0.67	0.75	0.71
022001	0	7.72	1.100	~0	1.54	1.03	1.10	1.08
021003	0	4.33	1.208	~0.003	1.80	1.28	1.54	1.42
023002	0.5%	6.17	1.158	~0.005	1.95	1.58	1.67	1.64
021004	0	3.52	1.277	~0.003	1.95	1.55	1.94	1.73
020005	0	2.81	1.368	~0.005	2.07	1.48	2.01	1.76
021002	0	6.17	1.192	~0.003	2.37	2.06	2.29	2.15
020004	0	3.55	1.333	~0.005	2.37	2.16	2.55	2.25
025003	0.1%	4.33	1.297	~0.005	2.57	2.46	2.74	2.62
021001	0	7.72	1.178	~0.003	2.75	2.64	2.85	2.76
020003	0	4.31	1.354	~0.005	3.05	3.11	3.33	3.25
025002	0.1%	6.17	1.306	~0.005	3.78	4.03	4.34	4.20
020002	0	6.11	1.362	~0.005	4.42	3.63	4.24	3.95
025001	0.1%	7.72	1.321	~0.005	4.95	3.48	4.30	4.01
020001	0	7.72	1.356	~0.005	5.49	3.18	4.17	3.65
020+02	0	9.36	1.369	~0.005	6.91	2.46	4.12	3.18
028002	1%	6.15	~1.37	~0.02	~4.6	3.27	4.08	3.72
028003	1%	4.24	~1.37	~0.02	~3.1	3.37	3.71	3.55
029002	5%	6.08	~1.6	~0.4(?)	~6.8	3.34	4.28	3.78
029003	5%	4.26	~1.6	~0.5(?)	~5.0	3.12	3.52	3.35

TABLE 1B

TARGET PARAMETERS AND EXTINCTION EFFICIENCIES
FOR 7-CYLINDER TOTALLY-REFLECTING ROUGH PARTICLES

Notation is the same as in Table 1A.

Target ID #	x_v	Q_{MIN}	Q_{MAX}	Q_{RAV}
030001	7.86	2.64	3.34	2.96
030002	6.19	2.26	3.21	2.93
030003	4.39	3.02	3.71	3.45
030004	3.62	3.17	3.65	3.44
030005	2.86	2.62	3.54	3.30
030007	1.86	3.44	4.24	3.83
030009	1.50	3.10	4.46	3.89

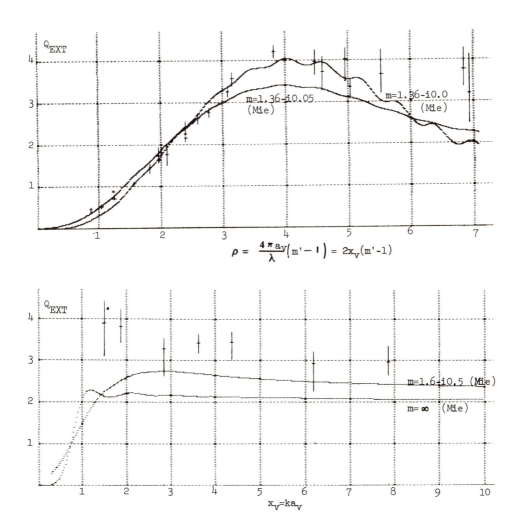

FIGS. 7A (top) and 7B (bottom).   Dependence of Q_{EXT} on ρ or x_V for 7-cylinder rough particles. Fig. 7A is for dielectric (penetrable) rough particles while Fig. 7B is for totally reflecting rough parti-cles. ρ is the phase shift parameter $= 2x_V(m'-1)$, and x_V is the size parameter of a smooth sphere having the same volume as a rough parti-cle.   The range of Q_{EXT} variation as each particle is rotated is shown by a vertical bar, and a horizontal mark on each gives the estimated Q_{EXT} averaged over random orientations.   In both figures, the Mie theory results for smooth spheres using appropriate refractive indexes are also shown by continuous curves.

The Q *vs.* ρ oscillation, which is characteristic of smooth particles, is here considerably damped, and Q_{RAV} becomes less dependent on ρ although Figure 7A suggests on the whole a gradual decrease of Q_{RAV} toward the asymptotic $Q_{RAV} = 2$ at $\rho = \infty$. Direct optical turbidity measurements on irregular quartz and diamond particles in liquid suspension (Hodkinson, 1963; Proctor and Barker, 1974) showed a similar rise in Q for small particle sizes. However, while their extinction curves also showed the damping of (the smooth sphere) oscillations after reaching a single peak in Q, the peak Q values were consistently lower than ours (even after accounting for their use of a G based on projected area) and occurred at considerably larger values of ρ(~6) compared to our value of $\rho \sim 4$.

A theoretical attempt to model d) by replacing the actual rough particle with an aggregate of independently scattering spheres (a sphere of the same mean radius plus spheres representing the pro-tuberances) could not explain the higher observed Q in this size range. However, the extinction signature by rough particles pre-sented in this report combined with some angular distribution data (Wang and Greenberg, 1978) and $\theta=180°$ empirical findings (Wang *et al.*, 1977) leads to the tentative conclusion that an irregular particle, near and above the first major resonance in size, casts more scattered light in the forward direction and less in the backward direction than an equivalently-sized smooth sphere.

The behavior of totally reflecting rough particles, as shown in the Q_{RAV} vs x_v plot of Figure 7B, is also curious. Here too Q_{RAV} is considerably higher than the corresponding smooth sphere as shown by the Mie curve for m=∞ in the same figure. The extinction peak for reflecting rough particles is shifted to a value of x_v that is smaller than that found for equivalent dielectric rough particles. (The same is true, of course, for smooth spheres). Furthermore, there seems to be no correlation between δQ and x_v.

3. CONCLUSION

The findings of this extinction study can be summarized as follows:

a) Judicious use of combined ray optics and Mie theory for spheres is capable of predicting the $\theta=0°$ scattering signa-tures for a variety of non-spherical/non-isotropic particles. This is particularly true for scatterers having fewer inhomo-geneities and/or interstices so that multiple reflections within the targets are insignificant.

b) For an aggregated particle, describing the scattering due to an individual component is by no means trivial. The effect of dependent scattering might be assessed if one replaces the complex near field by a simpler wave field consistent with physical reasoning. The most difficult case seems to be when the particle components are aligned along the incident direction.

Even for this case, a simple summation of independent scattering is not in serious error in the evaluation of extinction if the component separation exceeds about ten times the dimension of the largest component.

c) The fact that rough particles produce an extinction which is larger than that due to equivalent-volume spheres remains to be explained. But this empirical fact, along with the evidence that rough particles do not exhibit sphere-like oscillations in the extinction curve, is useful for estimating the extinction due to irregular particles.

Acknowledgements. The author would like to express his sincere appreciation to Prof. J. M. Greenberg and Dr. D. W. Schuerman for helpful discussions and for critically reviewing this manuscript. The analysis of these data is supported by the U. S. Army Research Office under contract No. DAAG29-79-C-0055.

REFERENCES

Born, M. and Wolf, E., 1965, "Principles of Optics," Pergamon, N.Y.

Bruning, J. H. and Lo, Y. T., 1971, *IEEE Trans. Ant. Prop.* AP-$\underline{19}$, 378.

Feenberg, E., 1932, *Phys. Rev.* $\underline{40}$, 40.

Greenberg, J. M., 1960, *J. Appl. Phys.* $\underline{31}$, 82.

Greenberg, J. M., Lind, A. C., Wang, R. T., and Libelo, L. F., 1967, *in* "Electromagnetic Scattering," L. Rowell & R. Stein, Eds., Gordon & Breach, N.Y., p. 3.

Greenberg, J. M. and Stoeckly, R., 1970, *in* IAU Symposium No. 36, L. Houziaux & H. E. Butler, Eds., D. Reidel, Dordrecht-Holland, p. 36.

Greenberg, J. M., Wang, R. T., and Bangs, L., 1971, *Nature, Phys. Sci.* $\underline{230}$, 110.

Hodkinson, J. R., 1963, *in* "Electromagnetic Scattering," M. Kerker, Ed., Pergamon Press, p. 87.

Liang, C. and Lo, Y. T., 1967, *Radio Science* $\underline{2}$, 1481.

Lind, A. C., Wang, R. T., and Greenberg, J. M., 1965, *Appl. Opt.* $\underline{4}$, 1555.

Lind, A. C., 1966, Ph.D. thesis, Rensselaer Polytechnic Inst., Troy, N.Y.

Montroll, E. W. and Greenberg, J. M., 1954, *in* "Proc. of Symposia on Appl. Math., Wave Motion and Vibration, Pittsburgh, Pa.," $\underline{5}$, McGraw Hill, N.Y., p. 103.

Proctor, T. D. and Barker, D., 1974, *Aerosol Science (GB)* $\underline{5}$, 91.

Rytov, S. M., 1955, *J. Eksp. Theor. Fiz.* $\underline{29}$, 605, (*Sov. Phys. JETP* $\underline{2}$, 466, 1956).

van de Hulst, H. C., 1946 "Thesis Utrecht," *Recherches Astron. Obs. d'Utrecht* $\underline{11}$, part 1.

van de Hulst, H. C., 1949, *Physica* $\underline{15}$, 740.

van de Hulst, H. C., 1957, "Light Scattering by Small Particles," Wiley, N.Y.

Wang, R. T., 1968, Ph.D. thesis, Rensselaer Polytechnic Inst., Troy, N.Y.
Wang, R. T., 1970, unpublished experimental results.
Wang, R. T. and Greenberg, J. M., 1976, *Appl. Opt.* <u>15</u>, 1212.
Wang, R. T., Detenbeck, R. W., Giovane, F., and Greenberg, J. M., 1977, Final report, NSF ATM75-15663 (June).
Wang, R. T. and Greenberg, J. M., 1978, Final report, NASA NSG 7353 (August).

SCATTERING BY PARTICLES OF NON-SPHERICAL SHAPE

R. H. Zerull, R. H. Giese, S. Schwill, K. Weiss

Ruhr Univ., Bereich Extraterrestrische Physik

Bochum, FRG

ABSTRACT

The scattering of particles deviating more or less from spherical shape is discussed based on microwave scattering measurements. To point out the differences between spheres and extremely irregular particles, fluffy structures will be of primary interest in this paper.

1.   GENERAL REMARKS

The results presented here are derived from microwave analog measurements. The principle of measurement as well as the equipment are described elsewhere (Zerull, 1976).

The scattering problem is described proceeding from a linear transformation (see van de Hulst, 1957)

$$\begin{Bmatrix} I_{s1} \\ I_{s2} \\ U_s \\ V_s \end{Bmatrix} = \frac{\lambda^2}{4\pi^2 \, r^2} \left| A_{ij} \right| \begin{Bmatrix} I_{o1} \\ I_{o2} \\ U_o \\ V_o \end{Bmatrix}.$$

Index o characterizes the incident, index s the scattered radiation; λ is the wavelength; r is the distance between the scattering particle and the observer. For the case where the incident light

273

is polarized perpendicular or parallel to the scattering plane and only the light polarized in the same direction is measured, the scattering event is described by the coefficients A_{11} and A_{22} of the scattering matrix alone. These coefficients are identical with the scattering functions i_1 and i_2, respectively:

$$I_{s1} = \frac{\lambda^2}{4\pi^2 r^2} i_1 I_{o1}$$

$$I_{s2} = \frac{\lambda^2}{4\pi^2 r^2} i_2 I_{o2} \quad .$$

For many applications the incident radiation is natural, *i.e.* unpolarized. In that case the depolarizing elements A_{12} and A_{21}, which are non-zero for non-spherical particles, must be taken into account. Separate measurements of these elements are published elsewhere (Zerull, 1976; Zerull *et al.*, 1978).

For natural light of intensity I_o the same formalism as above holds:

$$I_{o1} = I_{o2} = \frac{1}{2} I_o$$

$$i_1 = A_{11} + A_{12}$$

$$i_2 = A_{22} + A_{21}$$

$$i = i_1 + i_2$$

$$P = \frac{i_1 - i_2}{i_1 + i_2} \quad .$$

The total intensity i and the degree of linear polarization P as defined by the equations above will be used in the following sections to illustrate the scattering properties of irregular particles. They depend on the scattering angle θ and the material, size, and shape of the particles. The material is characterized by the complex refractive index m. The size is introduced by the size parameter x. This is referred to spheres and defined in this case as ratio of circumference and wavelength. If the parameter x is used to characterize the size of an irregular particle, it is always related to a sphere of the same volume, the so called "equivalent sphere". Although the volume may not always be a satisfactory base for comparison, it can be measured without difficulty. The choice of the same intermediate geometrical cross section as a base for comparison, as proposed and applied by several authors, is more meaningful only for sizes

$$x \gtrsim \frac{10}{|m-1|} \qquad \text{(Hodkinson, 1963)} ,$$

where the close connection between geometrical cross section G and the extinction cross section C_{ext}

$$C_{ext} = 2\ G$$

holds. Most of the particles treated in this paper, however, are smaller than required by the restriction mentioned above.

The particle size distribution for polydisperse mixtures was chosen according to a power law $dn = x^{-k}\ dx$ with $k = 2$. However, it is one of the advantages of microwave analog studies that all size distributions of interest can easily be calculated based on one set of measurements.

2.   RESULTS

2.1 Dielectric Particles

Different types of dielectric particles have been measured: rough spheres, cubes, octahedrons, convex particles, concave particles, and fluffy particles. Detailed results are published elsewhere (Zerull, 1976; Zerull *et al.*, 1974, 1977, 1978). To point out the general differences between scattering by spherical and by non-spherical particles as concisely as possible, cubes, convex and concave particles (see Fig. 3a), -- altogether 44 particles in the size range $1.9 \leq x \leq 17.8$ -- are compared to equivalent spheres in Figure 1. For larger scattering angles, the curves deviate strongly from one another. The intensity in the middle range of scattering angles is markedly higher for non-spherical particles and nearly isotropic up to the backscattering domain. Polarization properties are nearly neutral. Consequently, all resonance-like features, enhancement of backscattering, and strong negative polarization vanish in the case of non-spherical particles.

In Figure 2, a polydisperse mixture of fluffy dielectric particles (see Fig. 3b) in the size range $12.3 \leq x \leq 19.6$ is compared to equivalent spheres. Besides the features already mentioned above, in this case the absence of rainbow effects (based on the special geometry of spheres) and lower intensity in the forward direction are evident. The latter difference may be explained as follows: Although not really acting as individuals, the smaller constituents of the fluffy particles may preserve some of their individual features.

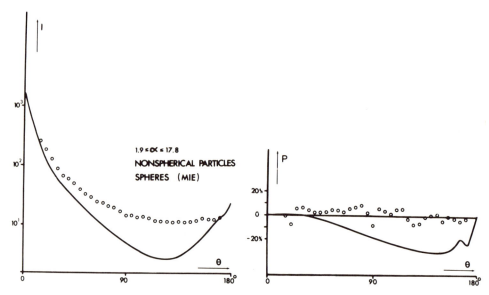

FIG. 1.   Scattering properties of a polydisperse mixture of non-
spherical particles (cubes, convex and concave particles) vs. equivalent spheres.

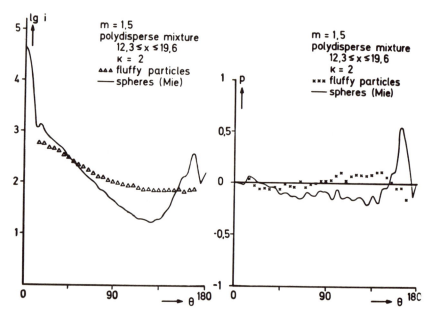

FIG. 2.   Scattering properties of a polydisperse mixture of dielectric fluffy particles vs. equivalent spheres.

The most important feature for this explanation is the ratio of
forward scattering to scattering at medium scattering angles, which
is by far lower in the case of smaller particles.

2.2 Absorbing Particles

Scattering of randomly oriented convex particles, which are
big enough for application of geometrical optics and whose absorption is strong enough to ensure that the refracted part of the
scattered radiation is totally absorbed, is easily determined by
diffraction and Fresnel reflection only. The scattering bodies discussed in this section do not satisfy the conditions mentioned above.
Figure 3 d-f shows three typical examples: irregular, longish, and
fluffy particles. The corresponding scattering properties are illustrated in Figure 4.

Most striking is the enhancement of the intensity level in all
cases of non-spherical particles. There are three reasons for this
feature:
1) Comparison to spheres is based on equal volume instead of
   equal cross section; this induces à factor of about 2 for
   the fluffy particles.
2) As already mentioned in the case of dielectric fluffy
   particles, smaller constituents of such particles may
   partially preserve their lower ratio of forward scattering
   to scattering at medium scattering angles even if they are a
   member of an aggregate; this would also increase the intensity at medium scattering angles.
3) The equivalent spheres absorb nearly all refracted radiation
   (illustrated by the close run of exact Mie curve and the
   curve considering Fresnel reflection only, outside the
   diffraction domain). Contrary to the case of non-spherical
   particles, especially of longish and fluffy shape, the refracted part of the radiation has a much better chance to
   reach the observer. This part, particularly dominant in
   the case of dielectric particles, is isotropic and unpolarized to a first approximation as pointed out in the case of
   dielectric particles. Thus, the increase of intensity due
   to refracted radiation should go hand in hand with a decrease
   in the degree of polarization, as is clearly observed. In
   the case of the fluffy particle and the longish particle
   the polarization maximum is decreased from nearly 100% (for
   the equivalent sphere) to about 50%. This means that about
   half of the radiation is unpolarized (*i.e.*, its origin is
   refraction). For this example, the intensity level is enhanced by a factor of about two again.
Besides the general increase of intensity and decrease of the polarization maximum, a further increase toward backscattering and shifting
of the location of the polarization maximum toward larger scattering

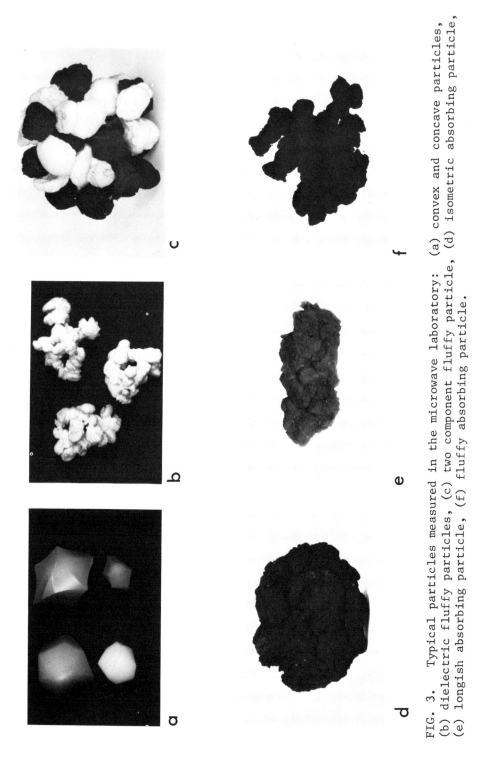

FIG. 3. Typical particles measured in the microwave laboratory: (a) convex and concave particles, (b) dielectric fluffy particles, (c) two component fluffy particle, (d) isometric absorbing particle, (e) longish absorbing particle, (f) fluffy absorbing particle.

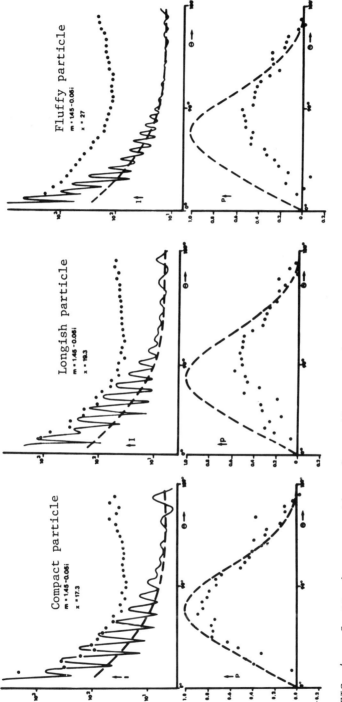

FIG. 4. Scattering properties of monodisperse mixtures of isometric, longish, and fluffy absorbing particles vx. equivalent spheres.

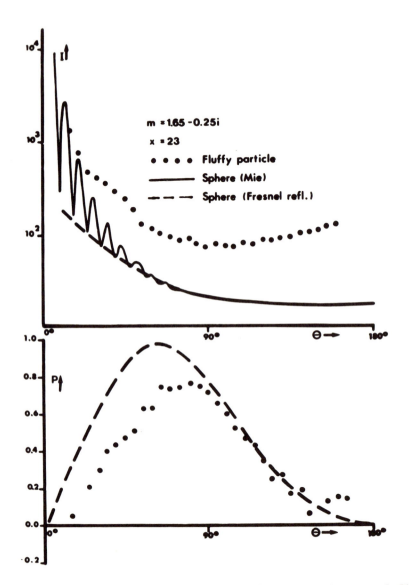

FIG. 5.   Scattering properties of a monodisperse mixture of fluffy
particles vs. equivalent spheres.

angles can also be observed. Both features also go hand in hand and
are sensitive to the shape of the particle. They are caused by
shadowing and multiple reflection effects as described elsewhere
(Giese *et. al.*, 1978; Wolff, 1975).

Figure 5 shows the scattering properties of a fluffy particle
of more strongly absorbing material. This example easily allows the
previous statements to be checked: due to stronger absorption there
is less increase of intensity; consequently, there is only a small
decrease of polarization. However, the enhancement of backscattering
and the shift of the polarization maximum (both of which are mainly
shape dependent) are fully preserved.

2.3 Two Component Fluffy Particles

Fluffy particles in general are not just an academic approach
to point out how strong scattering properties can differ from those
of spheres. Such shapes are found as constituents of the terrestrial
aerosol (Cheng, 1980, this volume), and they are also found among
particles of extraterrestrial origin (Brownlee, 1976). Most of the
fluffy particles collected by Brownlee consist of both dielectric and
absorbing constituents. Figure 3c shows one example of a set

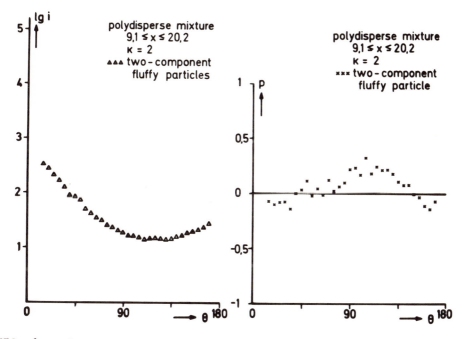

FIG. 6.   Scattering properties of a polydisperse mixture of two
component fluffy particles.

(9.1 ≤ x ≤ 20.2) which was built to simulate such extraterrestrial particles. The appropriate scattering properties are shown in Figure 6. They are not compared to any type of equivalent spheres. Readers of the field, however, may compare these results to empirical scattering functions derived from measurements of comets and zodiacal light and recognize the excellent correspondence.

REFERENCES

Brownlee, D. E., 1976, *Lecture Notes in Physics 48*, 279.
Giese, R. H., Weiss, K., Zerull, R. H., and Ono, T., 1978, *Astron. Astrophys.* *65*, 265-272.
Hodkinson, J. R., 1963, *in* "Electromagnetic Scattering," M. Kerker, Ed., Pergamon Press, 287.
van de Hulst, H. C., 1957, "Light Scattering by Small Particles," Wiley, N.Y.
Wolff, M., 1975, *Appl. Optics* *14*, 1395.
Zerull, R., Giese, R. H., 1974, *in* "Planets, Stars and Nebulae Studies with Photopolarimetry," T. Gehrels, Ed., Univ. of Ariz. Press.
Zerull, R. H., 1976, *Beiträge z. Physik der Atmosphäre* *449*, 168.
Zerull, R. H., Giese, R. H., Weiss, K., 1977, *Appl. Optics* *16*, 777.
Zerull, R. H., Giese, R. H., Weiss, K., 1977, *Optical Polarimetry*, SPIE Vol. 112, 191-199.

PHASE MATRIX MEASUREMENTS FOR ELECTROMAGNETIC

SCATTERING BY SPHERE AGGREGATES

Jerold R. Bottiger, Edward S. Fry, and Randall C. Thompson

Texas A&M University, Department of Physics

College Station, Texas   77843

ABSTRACT

An instrument which permits accurate measurement of all elements
of the phase matrix simultaneously has been constructed. Aggregates
consisting of attached polystyrene spheres are suspended in an
electric field and the phase matrix is measured.

1.   INTRODUCTION

During the development of a light scattering photometric polarimeter at our laboratory, numerous scattering matrix measurements
were made on aqueous suspensions of nearly monodisperse latex micro-
spheres. Small but significant discrepancies between measured and
calculated matrix elements were encountered, most notably in the 2,2
element, which is a sensitive indicator of sphericity. A microscopic
examination of the suspensions revealed aggregates of spheres, primarily doublets with decreasing numbers of triplets and higher order
aggregates, constituting in total about 10% of the scatterers. The
discrepancies were tentatively attributed to the presence of such
aggregates.

A study was undertaken to measure directly the scattering
matrices of these simple aggregates. The base spheres were 1091 nm
diameter (8 nm standard deviation) polystyrene spheres; the light
was from a He-Cd laser at wavelength 441.6 nm. Results for a doublet,
triplet, and quadruplet, as well as the single sphere, are presented
below.

2. THE INSTRUMENT

The light scattering photometric polarimeter has been described in detail elsewhere (Thompson *et al.*, 1977; Thompson, 1978). Briefly, the device uses four electro-optic modulators (EOM's) to modulate the state of polarization incident upon and scattered from a sample. Figure 1 shows the optical train employed. With a proper choice of the four modulation frequencies and amplitudes, there will exist in the Fourier spectrum of the detected intensity frequencies whose amplitudes are proportional to only one matrix element of the sample. Furthermore, there will exist at least one such frequency for every matrix element, so the entire matrix can be measured by monitoring amplitudes at appropriate frequencies using standard lock-in techniques. All of the matrix elements are individually measured, simultaneously and in real time, as the detector arm sweeps through scattering angles from ~0° to ~175°.

The dc component of the intensity is proportional to the 1,1 element (phase function) of the sample's scattering matrix. This is held to a constant value, defined to be unity, by a servo system which regulates the high voltage applied to the photomultiplier tube (Hunt and Huffman, 1973). Thus, we actually measure the normalized scattering matrix; the elements are restricted to values between ±1. The phase function itself may be determined either by opening the servo loop or by monitoring the high voltage required in the servo mode and calculating the intensity knowing the photomultiplier tube's response characteristics.

3. PARTICLE LEVITATION

To suspend a particle in air, an electrostatic levitation chamber was constructed along the lines described by Fletcher (1914) and Wyatt and Phillips (1972). Figure 2 is a cross-sectional schematic of the apparatus suspending a positively charged particle. Latex spheres, in a solution of water and methanol, are injected by an atomizer into

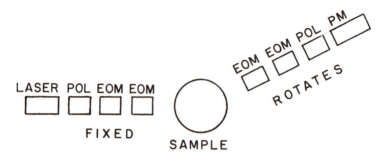

FIG. 1.   Schematic diagram of the photometric polarimeter.

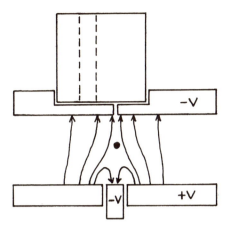

FIG. 2.   Schematic diagram of the particle levitation chamber.

the upper cylindrical chamber. The injection causes the particles
to become charged, with approximately equal numbers of positively
and negatively charged particles. When the cylinder is rotated to
the open position, particles fall through a small hole in the upper
plate into the levitation chamber.

The number of spheres in an aggregate is found by clocking its
rate of fall in zero electric field. Single polystyrene spheres of
1091 nm diameter were found to require 23 to 27 seconds to fall 1 mm;
doublets, 17 to 19 seconds; triplets, 13 to 15 seconds; and quad-
ruplets, 11 to 12 seconds. The spread in velocity for a particular
particle results from Brownian motion. It is difficult to distin-
guish among aggregates of more than five because of overlapping
velocity ranges.

Once a particle is selected, the top cylinder is rotated closed,
and the particle is held midway between the top and bottom plates by
manually adjusting the potential between the plates. The inhomogeneous field, which provides a radially inward force centering the
desired particle horizontally, also drives out all other particles
within a minute or two. Finally, a servo system is activated which
optically senses the particle's vertical position and adjusts the
plate potential as needed to keep the particle at a constant height.

Although trapped in a very small volume, the particle is still
subject to Brownian motion and rapidly changes its orientation. This
is clearly observed as a scintillation of the intensity, except in
the case of a single sphere where the intensity is quite steady. So,
while the sample is a single particle, the scattering matrix measured
is equivalent to that of a cloud of identical particles in many
orientations.

4.   RESULTS

    Figure 3 shows the 1,2 scattering matrix element for a singlet,
doublet, triplet, and quadruplet of 1091 nm spheres for scattering
angles from 12° to 165°. (This somewhat restricted range of
scattering angles was necessitated by the design of the electrostatic levitation chamber requiring very low levels of stray light).
The single particle Mie structure is clearly evident in the scattering
of the aggregates, although it becomes increasingly washed out as
the aggregate grows in complexity.   The same behavior is seen in
4,3 elements, Figure 4.

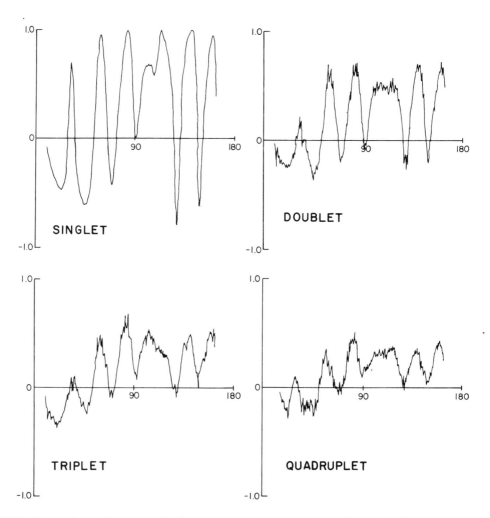

FIG. 3.   1,2 element of the scattering matrices for single and
aggregate spheres.

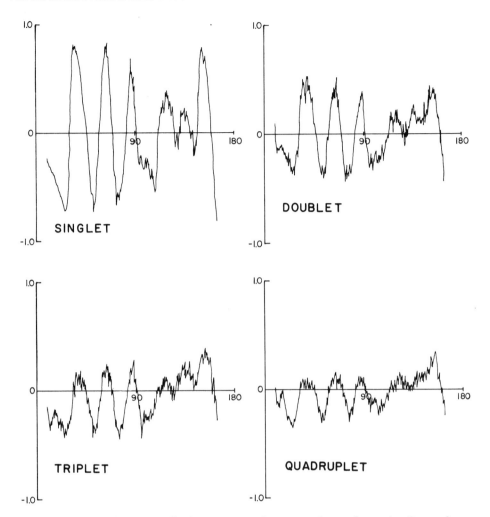

FIG. 4.   4,3 element of the scattering matrices for single and
aggregate spheres.

For all four samples, the 2,1 and (-)3,4 elements were identical
to the corresponding 1,2 and 4,3 elements and are not shown. The
eight elements which vanish for spheres (1,3; 1,4; 2,3; 2,4; 3,1;
3,2; 4,1; 4,2) were measured and found to be zero for all aggregates
as well as the single sphere. These also are not presented. The
vanishing of these eight elements for the aggregates is consistent
with the interpretation that we have in effect measured the particles
averaged over all orientations.

The 2,2 matrix element (Figure 5), which is unity at all
scattering angles for a particle with spherical symmetry, drops

strikingly in the case of aggregates of spheres. This element provides the most unequivocable evidence for the non-sphericity of the
aggregates.

The 3,3 and 4,4 matrix elements are presented in Figures 6 and 7.
They should be, and are found to be, the same for a single sphere.
However, even with random orientation there is no *a priori* reason
to expect them to agree in the case of aggregates. In fact, they are
not identical, as is most clearly seen in the backward scattering
angles of the quadruplet.

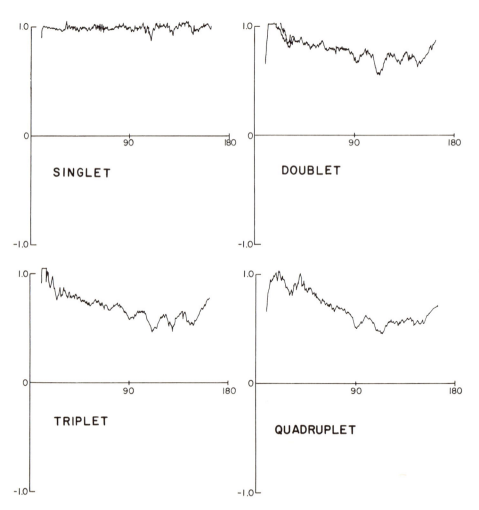

FIG. 5.   2,2 element of the scattering matrices for single and
aggregate spheres.

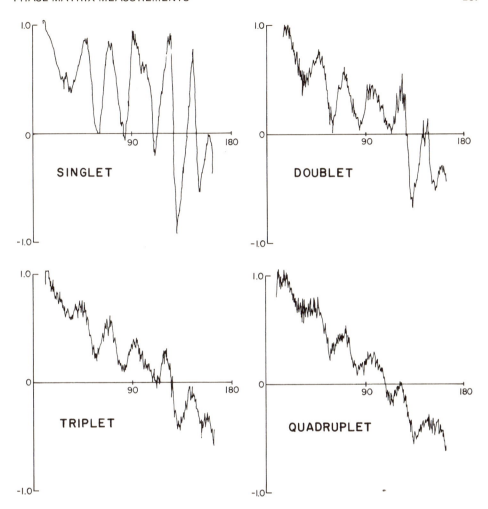

FIG. 6.   3,3 element of the scattering matrices for single and aggregate spheres.

5.   CONCLUSION

        Depending upon which matrix elements of the aggregates are examined, we can observe evidence of both the non-sphericity of the overall aggregate structure (*e.g.*, the 2,2 element) and the monodisperse spherical substructures. We feel it is important, especially when attempting to characterize irregular particles by their light scattering properties, that the entire scattering matrix be determined.

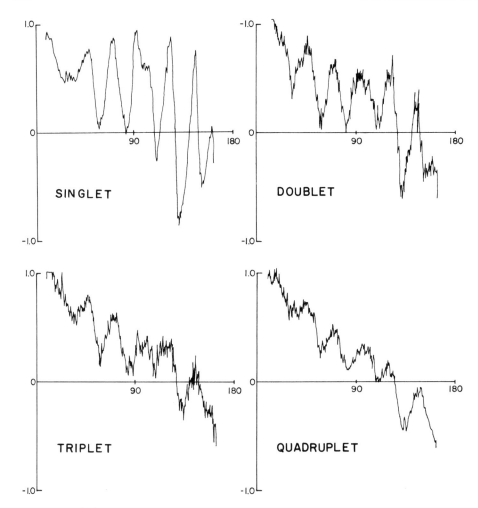

FIG. 7.   4,4 element of the scattering matrices for single and
aggregate spheres.

Acknowledgement. This work was supported by the Office of Naval
Research under grant #N00014-75-C-0537.

REFERENCES

Fletcher, H., 1914, *Phys. Rev.* <u>4</u>, 440.
Hunt, A. J., and Huffman, D. R., 1973, *Rev. Sci. Instru.* <u>44</u>, 1753.
Thompson, R. C., Fry, E. S., and Bottiger, J. R., 1977, *Proc. SPIE*
    <u>112</u>, 152.
Thompson, R. C., 1978, An Electro-Optic Light Scattering Photometric
    Polarimeter, Ph.D. thesis, Texas A&M Univ., College Station, Texas.
Wyatt, P. J. and Phillips, D. T., 1972, *J. Colloid Interface Sci.* <u>39</u>,
    125.

REFLECTIVITY OF SINGLE MICRON SIZE IRREGULARLY SHAPED DUST GRAINS

Madhu Srivastava

SUNY/Stony Brook, Department of Electrical Engineering

Stony Brook, New York 11794

D. E. Brownlee

University of Washington, Department of Astronomy

Seattle, Washington 98195

ABSTRACT

We have developed a microscopic technique to measure the optical albedo of *single*, micron size (5 to 7 µm in diameter) irregularly shaped dust grains in the wavelength range of $\lambda \sim .35$ µm to 0.6 µm. The experimental setup consists of a microscope, photometer, polarization analyzer and light sources. The illuminating spot size is limited to 5 µm in diameter. Dust grains are mounted on the tip of a thin (1 µm) glass fiber in order to reduce the background illumination. Our results on single, micron size carbonaceous chrondite type meteorites with application to the optical properties of interplanetary dust grains and stratospheric particles are presented. Our results indicate that single grains of carbonaceous chrondite material have an optical albedo of 5% at $\lambda \sim .35$ µm which decreases to 2% at $\lambda \sim 0.55$ µm. Scattered light from single grains is strongly polarized and it also exhibits an angular dependence.

In this communication, we report an experimental technique to measure the albedo of single micron size dust grains. This technique described here is suited to study the optical properties of interplanetary dust grains as well as other types of particles where one is forced to work only with single particles.

291

There are several commercial microscope setups available to measure the transmitted optical signal through micron size and sub-micron size biological and mineral specimens but none, to our knowledge, are used to measure absolute albedo of micron size irregular particles. Our experimental arrangement is shown in Figure 1. It consists of an interchangeable microscope objective for variable magnification and an optical photometer system mounted on top of microscope objective. The photometer system consists of a flip top mirror to view the image of the particle through an eyepiece. The size of the image seen through the eyepiece is the same as the one formed on the iris wheel, and there are eight different apertures available on the iris wheel. A magnification of 700 is needed for a grain size of 5 μm to completely fill the smallest aperture. On top of the iris wheel is the color filter wheel which has one aperature for white light and three color filters marked as U, B, V, a terminology commonly used in astronomy for ultraviolet,

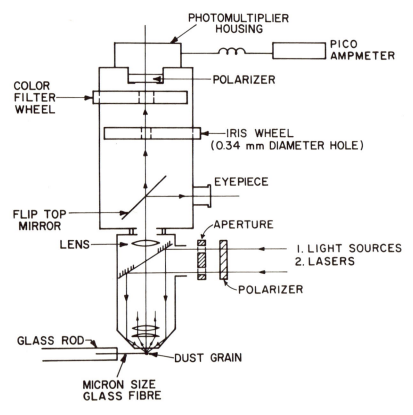

FIG. 1.   Experimental setup for measuring the albedo of single, micron size grains.   The optical arrangement is shown for dark field illumination.

blue and visible broad band color filters. The bandwidth of these
filters is approximately 700Å and the center wavelengths, λ_0, are:
U-3610Å, B-4440Å, V-5545Å.

Scattered light is detected by use of a photomultiplier. Polarization of scattered light can also be measured by the use of two
polarizers, one in the path of the incident light beam before illuminating the sample and the other in the path of the scattered light
in front of the photomultiplier. Reflectivity at wavelengths other
than U, B, V, can also be measured by use of monochromatic filters
in front of the light source. In our work, we used monochromatic
filters at $\lambda_0 \sim$ 5314Å, 6300Å and 6943Å. We also used a Helium-Neon
Laser at 6328Å as an illuminating source. Use of the laser gives
better control of the illuminating spot size, and detection of the
scattered light is made easier due to the high intensity of the laser
beams as compared to the intensity of incandescent light sources.

Single grains for reflectivity measurements are mounted on the
tip of a one micron size diameter glass fiber in order to reduce
scattering from background surfaces. We tried placing grains on a
glass microscope slide for albedo measurements but the reflection
from the glass surface introduced significant errors in the reflectivity measurements. The microscope system described here consists
of optical arrangements to illuminate the mounted grains in a bright
field or dark field. In a bright field, the incident light passes
through the objective lens vertically to illuminate the grain.
However, in the dark field arrangement, the incident light is reflected by mirrored surfaces (see Figure 1) to illuminate the mounted
grain at a grazing angle of 22° from the horizontal plane. In both
types of illumination, the incident light is not confined to one
specific angle but distributed over a definite solid angle as determined by the path of rays from the light source to the grain. For
bright field illumination the solid angle of the incident light beam
is much larger than in the dark field illumination. This difference
in solid angle of the incident beam is mainly due to the fact that
incident light passes through the objective lens for bright field
illumination but not for the dark field illumination. Scattered
light from the mounted grain is collected by the microscope objective
over a wide solid angle of approximately 145°. This solid angle of
scattered light depends upon the aperture size of the objective lens
and the distance of the grain from the objective lens.

The intensity of the light incident on the sample is measured
by use of a standard surface made out of Halon powder[1] coated on a
plastic or metallic surface. The absolute reflectance, ρ, of a Halon
coating of thicknesses greater than 6 mm is ρ = 0.991 in the

---

[1] Halon Powder is a trade name for the chemical polytetrafluroetylene
resin, sold by Allied Chemical, Specialty Chemicals Division, P.O.
Box 1087R, Morristown, N.J. 07960.

wavelength range of $\lambda \sim 0.35$ μm to 0.65 μm (Venable *et al.*, 1976).
Care was taken to avoid contamination of the Halon coating with
tobacco smoke or with metallic aluminum surfaces. We have found
that even slightest contamination of the Halon surface reduces its
absolute reflectivity by 10% or more. Calibration of the microscope
system was done by measuring the reflectivity of three standard
surfaces. One standard surface of size 3 cm × 1.5 cm was made by
painting several layers of gold paint on a glass microscope slide.
The average grain size of this painted surface was 1 μm. Similarly,
another standard surface was made out of black paint. A third
surface was made by polishing an aluminum plate with silicon carbide.
The plate was polished to a diffused surface with an average grain
size of 1 μm. We choose these three standard surfaces to cover the
reflectivity range of 4% to 70% in the wavelength range of $\lambda \sim 0.3$ μm
to 0.7 μm. The reflectivity of these standard surfaces was measured
by the microscope set up and by use of a spectrometer with an integrating sphere. It was found that deviation in reflectivity of standard surfaces as measured by these two separate methods was less than
one percent.

With the experimental arrangement as described above, we have
measured optical albedo of single micron size grains of several
materials. Here we report on the preliminary results of reflectivity
measurements of single particles of carbonaceous chondrite meteorites.

Figure 2 shows the reflectivity of single grains of carbonaceous
chondrite meteorite material. We studied several grains in the size
range of 5 to 7 μm. The albedo of these grains is the same with very
little deviation in reflectivity from one grain to the other. The
polarized component of the backscattered light in bright field illumination does not change significantly over the visible wavelength
range $\lambda \sim 0.35$ μm to 0.6 μm. However, the depolarized component
of the backscattered light changes drastically as evidenced by the
change in reflectivity of the grain from 6% at $\lambda \sim 0.35$ μm to less
than a percent at $\lambda \sim 0.55$ μm. Polarized and depolarized components
of the scattered light in dark field illumination exhibit a pattern
similar to each other. The polarized component has approximately
1% higher reflectivity than the depolarized one, and one observes
that the grain has a lower reflectivity at longer wavelengths which
rises in a power law fashion at shorter wavelengths.

We crushed several carbonaceous meterorite grains on a glass
surface to form an optically thick coating of crushed particles. On
a microscopic scale, the coating from crushed particles consists of
dark and bright patches. The dark patches presumably contain higher
carbonaceous contents or other opaque phases. The brighter patches
consist mainly of silicate materials. The average grain size of the
crushed particles is approximately 2 μm, and the average size of dark
and bright patches is approximately 10 μm. We have measured the reflectivity of these dark and bright patches, hereafter referred to as

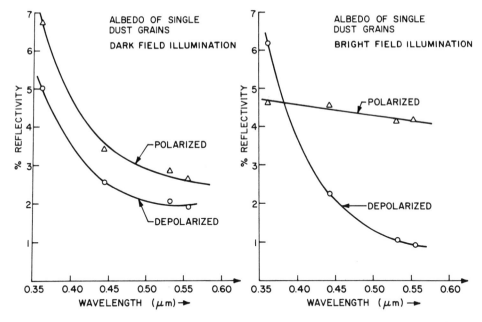

FIG. 2.   Reflectivity of single micron size grains.

carbonaceous material and silicate material of the meteorite particles, respectively. We find that wavelength dependent reflectivity
of the carbonaceous material of crushed meteorite particles is
similar to the reflectivity of the single 5 μm size meteorite grains
mounted on thin fibers (see Figure 3 for details). However, the reflectivity of the silicate materials exhibits a pattern in contrast
to the reflectivity of single grains. Figure 4 shows the reflectivity of silicate material as a function of wavelength. Polarized
arìd depolarized components of scattered light in bright field illumination give a lower value of reflectivity qf the silicate material
in contrast to the reflectivity measurement in dark field illumination. Another interesting feature to note is that the reflectivity
of silicate material is higher at longer wavelengths and decreases
at shorter wavelengths quite in opposition to the carbonaceous grains.
In latter cases the reflectivity is lower at longer wavelengths and
increases at shorter wavelengths.

    It has been found that the computed albedo of graphite grains
increases at shorter wavelengths (Wickramasinghe, 1973), a feature
similar to the measurements presented here. Computed data on silicate grains is not available, but the computed albedo of a mixture
of graphite, iron and silicate material decreases at shorter wavelengths which is similar to the wavelength dependent reflectivity

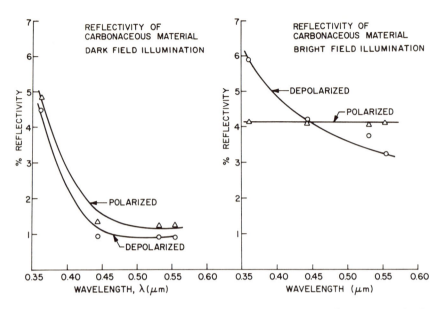

FIG. 3.   Reflectivity of carbonaceous contents of the crushed
meteorite particles.

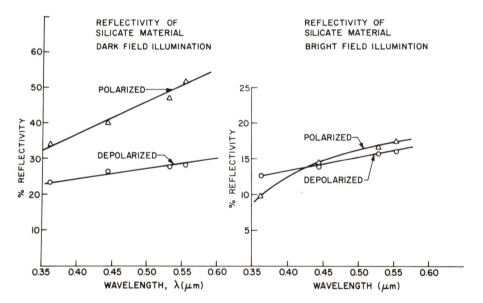

FIG. 4.   Reflectivity of silicate material of the crushed
meteorite particles.

of silicate material as reported here. The computed work of
Wickramasinghe (1973) does·not give any data on the polarization
of the scattered light. Therefore, the experimental results presented here can not be compared with albedo measurements, but the
wavelength dependence of the albedo can be compared with the computed
results.

The reflectivity of single grains (Figure 2) and the reflectivity of carbonaceous contents of crushed meteorite grains (Figure 3)
are similar. Reflectivity of single grains is the combination of
scattered light from carbonaceous and the silicate material of the
meteorite grain. The reflectivity of the silicate material is much
higher than the reflectivity of carbonaceous material, but scattered
photons from the silicate material will be absorbed by carbonaceous
material and thereby reduce the overall number of scattered photons
from a combined mixture of both constituents of meteorite particles.
We believe this could be the possible reason that the reflectivity
of single, micron size, carbonaceous meteorite grains is the same
as the reflectivity of the carbonaceous contents of crushed grains.

Reflectivity measurements of bulk samples of carbonaceous chondrite meteorites by Gaffey and McCord (1978) show that the reflectivity is approximately 4.5% in the visible region and falls to about
2.5% at $\lambda \sim 0.32$ µm. This drop in reflectivity of bulk samples at
shorter wavelengths is in contrast to our measurements of the reflectivity of single grains. In the microscope setup, the size of
the meteorite grain will influence significantly the amount of the
scattered light collected within a solid angle by the microscope
objective lens. The grain size will not alter the reflectivity
measurements of bulk samples where the scattered light is collected
in an integrating sphere over a large distance from the sample. The
effect of grain size on the reflectivity of single grains needs to
be examined in detail to find out the observed differences in the
reflectivity of single grains and bulk materials. Work is in progress
to measure the scattering functions of irregular shaped particles by
the use of laser light scattering techniques in order to study the
polarization of the scattered light and the effect of grain size on
the angle dependent scattering functions. Results will be published
in the near future.

Acknowledgements. We would like to thank Dr. John Adams of the
Geophysics Department for the use of his spectrometer to calibrate
the microscope setup for reflectivity measurements.

REFERENCES

Gaffey, M. J. and McCord, T. B., 1978, *Space Science Rev.* 21, 555.

Venable, Jr., W. H., Weider, V. R., and Hsia, J. J., 1976, Information Sheet on Optical Properties of Pressed Halon Coatings, July, 1976; Radiometric Physics Section (232-14), National Bureau of Standards, Washington, D.C.
Wickramasinghe, N. C., 1973, "Light Scattering Functions of Small Particles," Wiley, N.Y.

BIOLOGICAL PARTICLES AS IRREGULARLY SHAPED SCATTERERS

William S. Bickel

University of Arizona, Dept. of Physics and Microbiology

Tucson, Arizona   85721

Mary E. Stafford

University of Arizona, Dept. of Microbiology

Tucson, Arizona   85721

ABSTRACT

Light scattering techniques used in biology provide a fast, non-destructive probe that is very sensitive to small but significant biological changes that are *not seen by any other optical technique*. The S_{34} component of the scattering matrix is particularly sensitive to temporal changes in the biosystem as depicted in several examples.

Polarized light scattered from well-prepared biological particles such as spores, pollen, bacteria, and red and white blood cells in solution contains information specific to the biological condition of the particle. In addition, biological particles offer a wide range of geometrical shapes and internal structures that can be uniformly modified and controlled during light scattering (LS) experiments.

Our biological experiments (Bickel *et al.*, 1976; Bickel, 1980) constitute an active scattering system which involves remote sensing of light scattered from a very small volume element containing bioscatterers in a well-defined condition, illuminated by an incident beam prepared into an exactly known Stokes vector. In general, the

299

LS measurements which can be made on any system of regular, irregular or biological particles are motivated by the fact that:

1.   every system will scatter light,
2.   certain unique input-output Stokes vector combinations exist,
3.   the complete light scattering information is contained in a 4 x 4 sixteen element θ-dependent scattering matrix.

The measurement of all sixteen elements, even though they are not unique, offers a good test of the experimental apparatus and could turn up asymmetries caused by aligned systems. Light scattering measurements of any system can, therefore, be formally done, independent of the choice of a relevant scattering system or the desire to extract useful information from the signals.

Bioparticles as irregular scatterers are special. They live, grow, divide, and perform useful functions. A group of bioparticles, polydispersive with respect to every optical, geometrical, and electrical property, are still monodispersive in the fact that they perform a single biological function. However, we find that LS is able to probe a particular monodispersive property of a polydispersive system and give important phase information in the scattering matrix.

A quick scan through an elementary microbiology book will show many particles and particle properties of interest to scattering people:

1.   Cells can be perfect spheres, monodispersive in size, smooth in shape, and have "well-behaved" radial refractive index distributions.
2.   Some systems contain absorbing cores which can grow in size and shape.
3.   Internal and surface structures can be induced to change reversibly or irreversibly.

In general, the wide range of bioparticles gives a wide range of experimental conditions since they can have all geometrical shapes, varied internal and surface structures, and a wide range of refractive indices.

Since an irregular particle is not a fundamental system which can be "solved" in closed form to test fundamental theory, our experiments concentrated on the biological interpretation of the LS signals. Applied LS is a sensitive probe of biomaterial which gives information in addition to what can be learned from electron and light microscope, optical rotatory dispersion (ORD), circular dichroism (CD), and other optical measurements. Cells are complicated things. The structure and randomness of its organization that occurs at lower and lower levels of size - at higher and higher

magnification - is impressive. Biologists have already learned a
lot from light scattering's main competition, the microscope, which
forms an image of a Mie particle. The microscopist can look at the
top-bottom, left-right, front-back of a single cell and observe its
color. Laser scattering on the other hand sees the cell, probes the
internal and external structure in such a way that it scatters the
incident laser beam into all 4π space. What does the signal mean?
In contrast to the solar wind, comet tails, interstellar dust, blue
skies, red sunsets, and white clouds, phenomena caused naturally by
scattering, LS from biosystems is an artificially contrived phenomenon unnatural to biology and biological function. Therefore, the
usual optical and physical constants, refractive index (n), size
parameter [$r = r_o(1-A_n \cos \theta_n)$], and polarizability ($\alpha$), which predict
only how light will scatter from known geometrical structures, are
unimportant for the biological description of the system. Biological
structures and other irregular particles are too complex to be significantly characterized by such parameters. Just as there are no two
identical real irregular particles, there are no such things as two
identical bioparticles. Therefore, our signals are averaged over
many particles *of the same kind*.

In spite of these pessimistic criticisms of LS applied to
biology, LS has much to offer. It is a fast, nondestructive probe
that is very sensitive to small but significant biological changes
that are *not seen by any other optical technique*. Preparations of
biosystems for LS experiments require special care. In most cases,
preparing a biosample according to the stringent requirements de-
manded by LS is a research in itself. One must know (a) the history
of the particle from the time of extraction, (b) the media in which
it was grown, (c) all preparation techniques, and (d) its present
environment. There should be no mechanical destruction due to cen-
trifugation, straining, sorting, pouring, and mixing or chemical-
biodestruction due to growing, washing, and aging. In many cases,
bioparticles can be killed and fixed, making them safe and preserving
their geometrical and optical properties for a long period of time.
Finally, the LS sample should be (a) a well-defined biosystem free
of all particulate matter and other impurities, (b) alive, (c) bio-
significant, (d) reproducible, (e) durable enough for LS, (f) capable
of significant change, and (g) most important, of interest to someone.
Meaningful biological LS research requires the dedicated cooperation
of a microbiologist and a physicist. The microbiologist defines the
significant biosample, puts it into a well-defined biologically active
state, in a solution of known optical properties, and extracts from
the LS signals biologically significant results. The physicist
applies the technique, guarantees good data, and attempts physical
interpretation of the LS signals.

Because bioscatterers are very complex structures, we made no
attempt to invert the scattering signals to get sizes or structures.
The situation is not much different for other optical measurements

such as ORD and CD, long used by biologists, but incapable of yielding physical constants which uniquely characterize structure or function. Instead we proceed as follows:

1. Measure all matrix elements for a particular system and concentrate on the most sensitive ones.
2. Relate optical signals to structure and signal changes to biologically significant changes that have already been defined by other biological techniques.
3. Measure concurrent events such as LS signals and the action potential from a nerve axon (fiber) while it conducts its electrical signal.
4. Induce perturbations and transitions and monitor biochanges with LS signals.
5. Correct ORD and CD measurements for scattering effects.
6. Get signatures. These are LS signals which characterize a biosystem in its medium, in a particular state, prepared in a certain way. These can be compared to data obtained from chemical, immunological, electron and light microscopic, physical, and electrical measurements. Such signatures form the starting point (reference curves) for further study.

A collection of LS signals from thousands of varied samples into an atlas would have limited if any use because of their high dependence on the extreme complexity of the sample, their preparation, history, and biological state. However, each signature is an important reference curve with respect to further study. Fast non-destructive probing of the system's natural time development or its induced time development (before and after effects) will generate new signals. The signal changes can be elucidated by subtracting subsequent curves from the reference curve (or Rayleigh or Mie curve) and amplifying the difference. Other manipulations such as Fourier transforms, additions, subtractions, other mathematical combinations, measuring areas, number of inflection points, curve lengths, etc. can classify certain systems of complex particles even though the curve or procedure is not physically significant.

Our experiments with biological systems have concentrated on the S_{34} matrix element which connects $45°$ linear polarized light with circularly polarized light. It is especially sensitive to small structural changes. Circularly polarized light as applied to in ORD and CD measurements has always been an important probe for biomaterial. The S_{34} signal is especially model dependent, being the most difficult to fit to a theoretical curve. We attribute its sensitivity as a matrix element to the fact that it starts out as a zero signal for Rayleigh particles, but arises from the noise as the particle's features increase in size compared to the incident wavelength. Other signals such as S_{11}, S_{12}, S_{22}, S_{33}, S_{44}, already 100% for Rayleigh particles, mask small (few percent) deviations or changes

Our specific experiments have concentrated on time changes in S_{34} and in before-after situations. Figure 1 shows the noise level, stability, and reproducibility of the optical-electronic system and preparation techniques. Figure 2 shows the time development of the S_{34} matrix element signal from red blood cells. A plot of curve length as a function of time emphasizes the transition rate. Figure 3 is the complete scattering matrix obtained for three kinds of pollen in saline solution. The S_{43} and S_{34} signals are most sensitive to structural differences. Figure 4 shows signals from heat killed bacteria that have been grown in two different media. The nonzero difference curve at the bottom shows the importance of growth media, the richer media (BHI) produces more DNA. Figure 5 shows the signals from a normal and mutant bacteria. In this case, the flagellated mutant (typhi O) has had its flagella removed. The difference curve is at the bottom. These data demonstrate the sensitivity of the S_{34} matrix element to biological structure and the care with which samples must be defined and prepared.

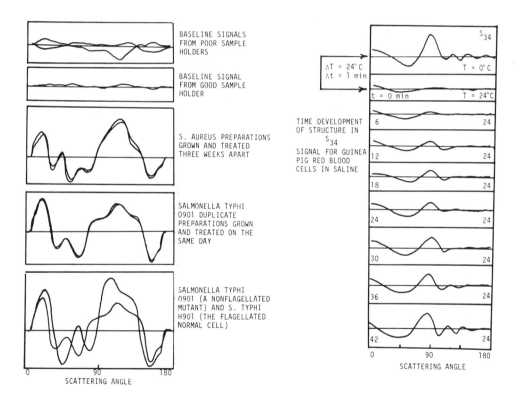

FIG. 1.   Noise level, stability, and reproducibility of both the optical-electronic system and the preparation techniques.

FIG. 2.   The time history of the S_{34} component of the scattering matrix from red blood cells.

W. S. BICKEL AND M. E. STAFFORD

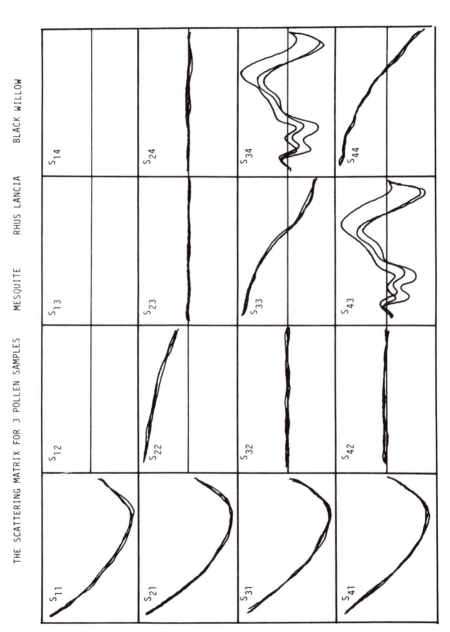

FIG. 3. The complete scattering matrix obtained from three kinds of pollen (Mesquite, Rhus Lancia, Black Willow) in saline solution.

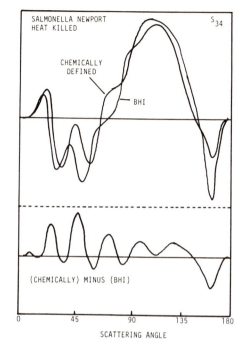

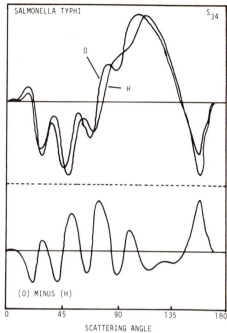

FIG. 4.   Scattered light signals
from heat killed bacteria that
have been grown in two different
media.

FIG. 5   Signals from normal (H)
and mutant (O) bacteria.

Acknowledgements.   We thank the donors of the Petroleum Research
Fund, administered by the American Chemical Society, for partial
support of this research.   Support from NSF is also acknowledged.

REFERENCES

Bickel, W. S., Davidson, J. F., Huffman, D. R., and Kilkson, R.,
    1976, *Proc. Nat. Acad. Sci. 73*, 486.
Bickel, W. S., 1979, *Applied Optics 18*, 1707.

INFERENCE OF SCATTERER SIZE DISTRIBUTION

FROM SINGLE SCATTERING MATRIX DATA

William A. Pearce

EG&G Washington Analytical Services Center, Inc.

Riverdale, Maryland   20840

ABSTRACT

We have studied procedures for inferring the scatterer size distribution from the polarization of scatter radiation. The information content of specific matrix element measurements is evaluated with assessments of the impact of the assumption of sphericity and other modeling errors as well as random measurement errors on the computed size distribution.

1.   INTRODUCTION

The problem of inferring the size distribution of light scattering particles from single scattering experimental data is functionally linear provided the particle shape and refractive index are known. Even when these are known *a priori* and with accuracy, limitations on our inferred knowledge of the size distribution exist which are traceable to incompleteness of the experimental data and to the finite precision of the data. Our immediate goals are to detect inadequacies in any model assumptions and to compute an estimate of the size distribution along with estimates of probable error bounds for the distribution arising from experimental errors.

We start with the (linear) equations relating a set of experimental observations X_j to the size distribution $n(r)$ and the monodisperse matrix elements $\hat{S}_j(r)$.

$$X_j = \int \hat{S}_j(r)\ n(r)\ dr \qquad (1)$$

307

where the index j encompasses matrix element designation, wavelength,
scattering angle, etc. It is convenient to solve these equations by
expressing the size distribution n(r) as a linear combination of known
functions and solving the resulting set of algebraic equations for
estimates of the coefficients of the known basis functions. We thus
write

$$n(r) = \sum_{\ell} B_{\ell}(r)\, \eta_{\ell} \tag{2}$$

so that equation (1) becomes

$$X_j = \sum_{\ell} \hat{S}_j(r)\, B_{\ell}(r)\, dr\, \eta_{\ell} \tag{3}$$

or $X_j \equiv \sum_{\ell} S_{j\ell}\, \eta_{\ell}$ or in matrix form $\underset{(m\times 1)}{\underline{X}} = \underset{(m\times n)}{\underline{\underline{S}}}\ \underset{(n\times 1)}{\eta}$. (3a)

Equations (3a) are in general not well conditioned. Even though the
number of experimental observables X_j may be far greater than the
number of parameters (η_{ℓ}), the data may not provide sufficiently
independent information.

2. BASIS FUNCTIONS

The choice of basis functions is important since it controls
both the flexibility of the model space and the smoothness of allowed
solutions (and, hence, the influence of poor conditioning). We are
faced with a tradeoff which is fundamental to inversion problems of
this type. We have chosen to use the B-Splines whose properties
were described by Curry and Schoenberg (1966).

Let $\ldots \leq X_{-2} \leq X_{-1} \leq X_0 \leq X_1 \leq \ldots \leq X_n \ldots$ be a nondecreasing
sequence of real numbers. A general spline function $S_{(n)}(X)$ of order
n (degree n-1) having these knots is defined as belonging to the class
C^{n-2} such that in each interval $(X_{\nu}, X_{\nu+1})$ it reduces to a polynomial
of degree not exceeding n-1. A B-Spline is a spline with a minimal
number of knots, namely n+1 (and whose knots are restricted by
$X_{\nu} < X_{\nu+n}$). Splines include the class of polynomials, and B-Splines
form a basis for splines. If we denote the B-Splines of order n
(with knots $X_{\nu}, \ldots, X_{\nu+n}$) by

$$B_{\nu}(X) = B_{(n)}(X\,;\, X_{\nu}\,,\, X_{\nu+1}, \ldots, X_{\nu+n})$$

then the functions $B_{-n+1}(X)$, $B_{-n+2}(X),\ldots, B_0(X)$ are linearly independent in the interval (X_0,X_1) and, in this interval, form a basis
for π_{n-1} (the class of polynomials of degree not exceeding n-1).

The character of the B-Splines is controlled by the knots. For
example, when two knots coalesce, the continuity class of a B-Spline
having these knots reduces by one in the neighborhood.

B-Splines are proper frequency functions since they are non-
negative, are zero outside the range of their knots, and are normalized so that their integral over the real line is unity. This makes
them especially attractive for our purposes. By adjusting the under-
lying knot distribution we may seek the most economical representation of the size distribution and largely avoid requiring detailed
cancellations to achieve a good fit. Changing the order of the B-
Splines modifies their smoothness properties.

We calculate the B-Splines using de Boor's (1977) recurrence
formula

$$B_{(n)}(X;X_i,\ldots,X_{i+n}) = \frac{n}{n-1}\ \frac{X-X_i}{X_{i+n}-X_i}\ B_{(n-1)}(X;X_i,\ldots,X_{i+n-1})$$

$$+\ \frac{X_{i+n}-X}{X_{i+n}-X_i}\ B_{(n-1)}(X;X_{i+1},\ldots,X_{i+n})$$

with the sequence initialized using

$$B_{(1)}(X;\ X_1,X_2)\ \begin{cases} = 0 \text{ for } X < X_1 \\[2mm] = \dfrac{1}{X_2-X_1} \quad \text{for } X_1 < X < X_2 \\[2mm] = 0 \text{ for } X > X_2\ \ . \end{cases}$$

The common practice of inverting by approximating the size distribution by a constant over increments in r is precisely the same as
using B-Splines of order 1. Since the basis functions in this case
provide no smoothing to dampen anomalies introduced by ill-conditioning, auxiliary smoothing must be used (see for example, Shaw, 1979).
By using a higher order basis, we can relinquish reliance on auxiliary
smoothing.

3. SOLUTIONS

To solve the overdetermined set of equations (3a) we decompose the matrix S using a singular value decomposition (Golub and Reinsch, 1970)

$$S = U \; \Sigma \; \tilde{V} \tag{4}$$

where $\tilde{U}U = \tilde{V}V = \tilde{V}V = I_n$, $\Sigma = \text{diag}(\sigma_1,\dots,\sigma_n)$, $\tilde{V}$ denotes the transpose of V, and I_n is the n-dimensional unit matrix. The columns of the matrix V(U) are the n orthogonalized eigenvectors associated with the n (n largest) eigen values of $\tilde{S}S(S\tilde{S})$. The diagonal elements of Σ are the non-negative square roots of the eigen values of $\tilde{S}S$ (they are called singular values).

Since S is not a square matrix, it possesses no true inverse, but instead has a pseudo inverse S^+ which can be constructed easily: $S^+ = V \; \Sigma^+ \; \tilde{U}$ where $\Sigma^+ = \text{diag}(\sigma_i{}^+)$ and

$$\sigma_i{}^+ = \begin{cases} 1/\sigma_i & \text{for } \sigma_i > 0 \\[2mm] 0 & \text{for } \sigma_i = 0 \;. \end{cases}$$

Thus, we obtain as a solution $V \; \Sigma^+ \; \tilde{U} \; X = \eta$ \hfill (5)

It is important to notice that the inversion acts only on that part of the data which is not orthogonal to $\tilde{U}$. That is, we effectively *filter* the data. Suppose we write the data as

$$X = Ut + \varepsilon \tag{6}$$

where $\tilde{U}\varepsilon = 0$. Now $\tilde{U}UX = \tilde{U}UUt + \tilde{U}U\varepsilon = Ut$, so that

$$(X - U\tilde{U}X) = \varepsilon \quad \text{or} \quad (I_m - U\tilde{U})X = \varepsilon \;. \tag{7}$$

(Remember that $U\tilde{U} \neq I_m$).

The elements which are incorporated into the filter are essentially the model assumptions; namely particle shape, orientation distribution, refractive index, and the flexibility and suitability of the basis functions B. We thus have a direct test of whether the model assumptions are capable of an accurate description of the data.

Instabilities in the inversion enter primarily through amplification of small errors by the Σ^+ matrix which is composed of the inverses of the singular values. Drastically small singular values tend to amplify errors.

We can, in addition, examine the propagation of random Gaussian errors into the solution. Remembering that equation (5) is essentially a linear transformation of the data, we can construct K_η, the covariance matrix of the η's, given the covariance matrix of the data, K_x:

$$K_\eta = (V \ \Sigma^+ \ \tilde{U}) \ K_x \ (U \ \Sigma^+ \ \tilde{V})$$

Note that if the random experimental error is Gaussian, the induced random error in the basis function coefficients (the η's) will be also. We can go further to obtain the covariance matrix of the solution noting that

$$n(r_i) = n_i = \sum_j B_j(r_i) \ n_j = \sum_j B_{ij} \ n_j = \underline{B} \ \underline{n}$$

and using the covariance of the η_ℓ's:

$$K_{n(r)} = B(V \ \Sigma^+ \ \tilde{U}) \ K_x (U \ \Sigma^+ \ \tilde{V}) \ \tilde{B}$$

4.   EXAMPLE

As an example to illustrate the procedures, we have chosen a bimodal size distribution $n(r) \sim n_1(r) + n_2(r)$ with each component given by a lognormal form

$$n_i(r) = w_i \ \frac{d_i}{(r-c_i)} \ \exp \left\{ -\left[a_i \ \ell n \left(\frac{r-c_i}{b_i - c_i} \right) \right]^2 \right\} \quad ; r > c_i$$

$$= 0, \quad r < c_i$$

$$= 0, \quad r > 2.0$$

with parameter values given by

i	a_i	b_i	c_i	d_i	w_i
1	0.896	0.45	0.05	0.2528	0.5
2	0.896	1.3	0.75	0.2528	0.5

Particles were taken to be spherical (so that $\hat{S}$ could be computed
with a standard Mie code), and the refractive index was m = 1.55 + 0i.
The assumed data set consisted of measurements of S_{11} and S_{12} at two
wavelengths (λ = 340 and 540 nm) and at 69 angles: θ = 0°(.2)3°,
3°(1.0)10°, 10°(2.0)30°, 30°(5.0)160°, 160°(2.0)180°. Normalization
was provided by extinction data.

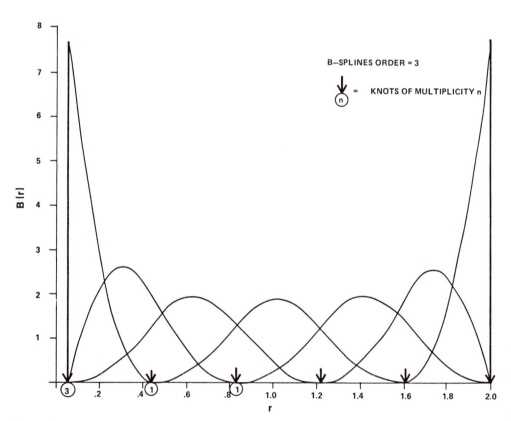

FIG. 1.   The set of B–Splines used in the present inversion example.

For 6 knots equidistantly spaced at r = 0.05 (0.39) 2.0, the RMS
error of the filtered data was minimized (RMS error = 0.24) with
B-Splines of order 3. These are shown in Figure 1. Triple knots at
0.05 and 2.0 were assumed to admit the full set of B-Splines. The
singular values were:

Eigenvector No.   1   2   3   4   5   6   7

Singular Value .157(6) .793(4) .141(4) .744(3) .223(3) .439(1) .760(2)

where the parenthetical numbers denote powers of 10; e.g., .157(6) =
$.157 \times 10^6$. The radial eigenfunctions are shown in Figure 2.

Figures 3 and 4 illustrate the behavior of the eigenvectors of
$S\tilde{S}$ for portions of the 340 nm data (S_{11} and S_{12}). Remembering that
these act as filters, the regions where the various eigenvectors are
large indicate experimental regions of sensitivity. By comparing with
Figure 2 we can relate these to certain regions in r-space. For
example, eigenvectors 1 and 2 which are most important in the vicinity of the diffraction peak at very forward angles are related to
the large-r region of the size distribution, while the innermost
regions of the size distribution are more sensitive to measurements
of S_{11} and S_{12} between 20 and 60 degrees. Figure 4 also shows the
S_{12} data and filtered data at 340nm (this is the data region of worst
fit).

The deduced and true size distributions are compared in Figure 5.
The error bars shown in the figure correspond to data with 1% noise;
i.e., $\sigma_x^2 = (.01\ x)^2$, K_x is assumed diagonal (data are independent).
As may be seen, the equidistantly spaced order 3 B-Splines are not
quite flexible enough to fit the central minimum precisely. Adjustments to the precise placement of the knots would likely provide a
better fit.

5. CONCLUSIONS

When the modeling assumptions which underlie the inversion
process are particularly crucial or are uncertain, as is likely to
be the case when non-spherical particles are involved, the singular
value decomposition approach offers valuable insights into the
adequacy of the model. B-Splines, used as the basis functions, offer
a high degree of flexibility, as well as adjustable smoothness properties which are useful for damping instabilities induced by poor
conditioning.

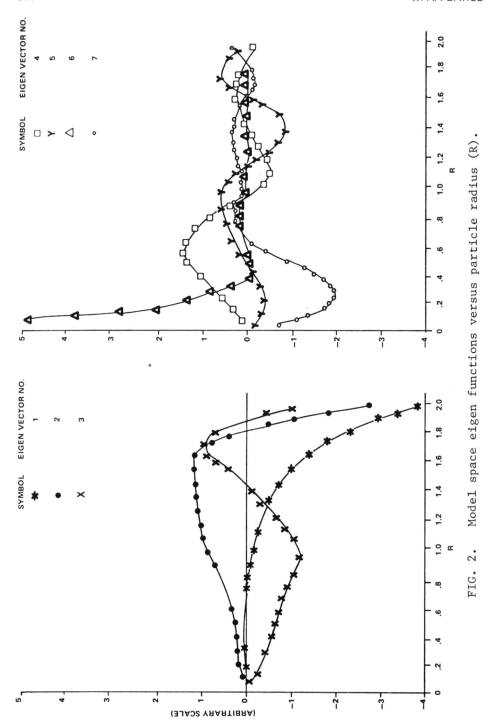

FIG. 2.   Model space eigen functions versus particle radius (R).

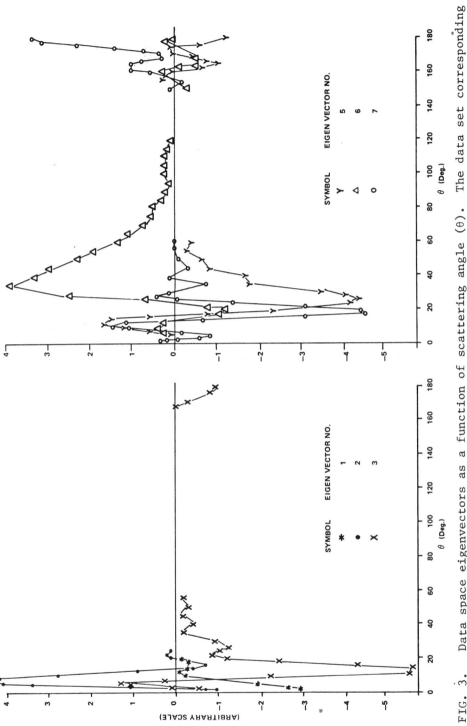

FIG. 3.   Data space eigenvectors as a function of scattering angle (θ).   The data set corresponding to this figure consists of measurements of S_{11} at a wavelength of 340 nm.

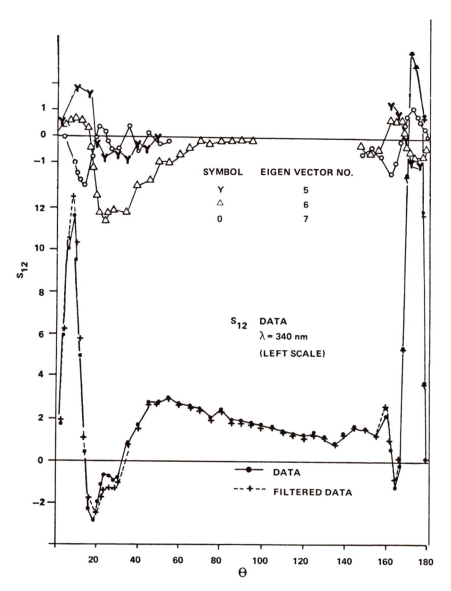

FIG. 4.   A comparison of the most significant data space eigenvectors
for S_{12} data with data and filtered data at a wavelength 340 nm.

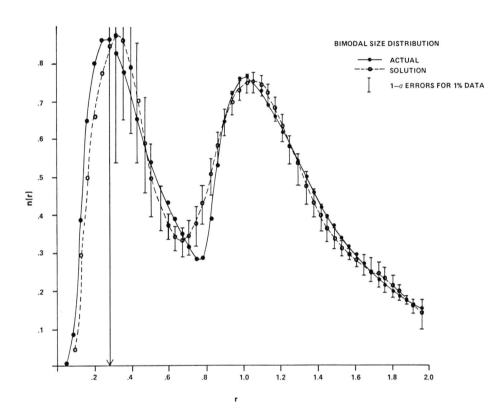

FIG. 5.   A comparison of the actual size distribution (——•——) with the inverted solution (--•--). The error bars reflect the statistical impact (1 st. dev.) of a 1% random error in the input data.

Acknowledgements.   I want especially to thank A. C. Holland for his support, encouragement and advice. This work was supported by NASA Contract NAS 6-2610.

Question (WANG):   In your technique, do you have to predetermine the refractive index and shape of the particle?

Answer (PEARCE):   In the particular case I've treated here, they were assumed known.   For a realistic application, the *proper* set of Ŝ's (which contain the refractive index, particle shape and/or shape distribution information) would have to be determined by an interactive scheme or by some other means.

REFERENCES

Curry, H. B., and Schoenberg, I. J., 1966, *J. d' Anal. Math.* <u>17</u>, 71.
de Boor, D., 1977, *SIAM J. Numer Anal.* <u>14</u>, 441.
Golub, G. H., and Reinsch, C., 1970, *Numer. Math* <u>14</u>, 403.
Shaw, G. E., 1979, *Applied Optics* <u>78</u>, 988.

LIGHT SCATTERING BY IRREGULARLY SHAPED PARTICLES VERSUS SPHERES:
WHAT ARE SOME OF THE PROBLEMS PRESENTED IN REMOTE SENSING
OF ATMOSPHERIC AEROSOLS?

John A. Reagan and Benjamin M. Herman

University of Arizona

Tucson, Arizona, 85721

ABSTRACT

Effects of light scattering by irregularly shaped particles
versus spheres are discussed in the context of defining how they influence aerosol properties inferred from optical remote sensing
measurements. Particular attention is given to identifying effects
on aerosol properties inferred from data acquired by monostatic and
bistatic lidar and spectral solar radiometer measurements.

1. INTRODUCTION

Remote sensing determinations of the atmospheric aerosol properties required for atmospheric radiative transfer studies ($i.e.$,
particle size distribution, refractive index, and vertical number
density profile) are extremely difficult to make even for the idealized condition of spherical particles. Factors such as spatial inhomogeneities in particle size distribution and refractive index, as
well as variation in particle refractive index with particle size,
may prevent making such remote sensing determinations and possibly
overshadow any effects due to particle nonsphericity. The relative
importance of each of these factors, including particle nonsphericity,
is obviously difficult to assess. The approach taken by the University of Arizona Aerosol and Radiation Research Group in attempting
to characterize the radiative properties of atmospheric aerosols has
been to assume that the particles may be treated as equivalent spheres
of constant refractive index. Assessment of the validity of this
approach is made on the basis of how well the results obtained by
various remote sensing techniques agree or disagree with one another.
This paper briefly discusses:

319

1) the remote sensing procedures we have employed to determine
   various aerosol properties subject to these assumptions,
2) how these determinations may be biased by the presence of
   non-spherical particles, and
3) how important are these biases compared to other uncertainties associated with the overall remote sensing problem.

2. MEASUREMENT TECHNIQUES AND DATA ANALYSIS PROCEDURES

The different measurement techniques employed in our aerosol and
radiation experiments are listed below along with brief summaries of
the analysis procedures applied to and the aerosol properties sought
from the data obtained with each technique.

2.1 Solar Radiometer

Multi-wavelength solar radiometer measurements are made at
visible and near infrared wavelengths with a filterwheel radiometer
to determine aerosol optical depths at different wavelengths as described by Shaw *et al.* (1973). The optical depth data are inverted to
determine the columnar or height integrated aerosol size distribution
by the inversion procedure described by King *et al.* (1978). This
procedure has already been employed to extract aerosol size distributions from more than 75 days of radiometer observations. A particularly attractive feature of the aerosol optical depth measurements
is that they are relatively insensitive to the refractive index of
the aerosol particles (at least for the index range thought to be
representative of typical atmospheric aerosols). Accurate size distribution inversions may be obtained without exact *a priori* knowledge
of the true particle refractive index (King *et al.*, 1978).

2.2 Monostatic Lidar

Backscatter measurements are made along different slant paths
(i.e., at various zenith angles) with a calibrated ruby lidar system.
The measurements are processed by a multi-angle integral solution of
the lidar equation (Spinhirne, 1977) to extract vertical profiles of
the aerosol extinction coefficient and the extinction to backscatter
ratio, S. The solution approach assumes that S is constant with
height through layers of optical thickness ~ 0.05 (which typically
includes the entire mixing layer for the relatively clear atmospheric
conditions representative of Tucson). Some degree of horizontal
homogeneity (backscatter variations of 10 to 20% can easily be
tolerated) is also required for accurate implementation of the
solution technique. The constraint that S be constant with height
requires that the refractive index and the shape of the size distribution of the aerosol particles remain constant with height.

However, no *a priori* assumptions regarding particle shape, refractive index and size distribution shape or the value of S are required to implement the solution technique. The error levels in aerosol extinction and S derived from the solution procedure reflect whether or not the requirements of horizontal homogeneity and S remaining constant with height are reasonably well met for a given set of slant path measurements. Results from many days of observations reveal that the aerosol extinction coefficient and S value can frequently be determined with an *rms* uncertainty between 10 and 20% (Spinhirne, 1977; Spinhirne *et al.*, 1979).

2.3 Bistatic Lidar

Angular scattering measurements of the I_1 and I_r Stokes components* of light scattered by a remote volume of air are made with a calibrated ruby bistatic lidar system. By simultaneously varying the elevation angles of the lidar transmitter and bistatic receiver, scattering measurements are made for scattering angles typically between 120° to 160° and a fixed height above ground (typically about 1 km). Horizontal inhomogeneities, to the extent that they are due to aerosol number density variations, are scaled out through the use of simultaneous slant path backscatter measurements. For typical operating conditions in Tucson, we have found it possible to determine the absolute values of the aerosol contributions to I_1 and I_r with an *rms* uncertainty between 5 to 10%.

The I_1 and I_r aerosol angular scattering measurements are used in combination with inferences of the aerosol extinction to backscatter ratio, S, and columnar aerosol size distribution, obtained by the methods mentioned in 2.1 and 2.2, to determine the mean aerosol particle refractive index (both real and imaginary components) by the technique described by Reagan *et al.*(1977). Using the refractive index value determined by this technique, the I_1 and I_r aerosol angular scattering data are then inverted to determine a unit volume aerosol size distribution which, as a consistency check, is compared to the columnar size distribution already obtained by inverting the multi-wavelength aerosol optical depth data.

2.4 Spectral Flux Sensor

Multi-wavelength total and diffuse hemispheric fluxes are measured with a spectral flux sensor. The measured fluxes serve as reference values which are compared with theoretically computed

---

* The l and r subscripts denote intensities for the electric field components parallel and perpendicular, respectively, to the scattering plane.

fluxes which are calculated using the aerosol size distribution, mean
refractive index, and vertical number density profile inferred by the
optical remote sensing techniques mentioned in 2.1, 2.2, and 2.3.   In
addition, the spectral diffuse-direct ratios of the measured fluxes
are analyzed by the method described by King and Herman (1979) to
infer the aerosol imaginary refractive index component and ground
albedo.

3.   EXPECTED BIASES DUE TO IRREGULARLY SHAPED PARTICLES

Of the various remote sensing measurements which we make, the
multi-wavelength optical depth measurements should be the least
affected by non-spherical particles.   Measurements of light scattering
by numerous types of irregular particles (*e.g.*, Hodkinson, 1963;
Holland and Draper, 1967; Holland and Gagne, 1970; Zerull and Giese,
1974) indicate that scattering in the near forward direction is
similar to that of spheres.   As noted by Pollack and Cuzzi (1978),
the extinction efficiency of polydispersions of randomly oriented
irregular particles can be enhanced over that of spheres of equivalent volume by 20 to 30% for particles with size parameter* values
greater than about 4.   This discrepancy can apparently be largely
scaled out by normalizing to spheres of equivalent surface area
rather than equivalent volume.

Angular scattering measurements of the intensity and polarization characteristics of monodisperse and relatively narrow polydisperse collections of randomly oriented irregular particles clearly
reveal large departures from the scattering properties of spheres
(whether normalized to equivalent volume or area) for scattering
angles greater than about 40° (Holland and Gagne, 1970; Zerull and
Giese, 1974; Pinnick *et al.*, 1976; Perry *et al.*, 1978).   For
scattering angles around 120°, the I_1 and I_r scattered Stokes components for irregular particles generally exceed those of spheres
of equivalent volume, sometimes by factors of 5 or greater.   However,
these components are generally lower than those of spheres of equivalent volume for scattering angles close to 180°.   Based on the
semi-empirical phase function curves derived by Pollack and Cuzzi
(1978) to fit some of the scattering measurements of Zerull and Giese
(1974), it appears that the backscatter coefficient for light
scattering from polydispersions of randomly oriented irregular particles can be 3 to 10 times smaller than that of spheres of equivalent volume.   As the extinction efficiency of irregular particles
is slightly greater than that of spheres, this indicates that the
extinction to backscatter ratio, S, for irregular particles can also

----

* The size parameter is the particle circumference to wavelength
  ratio.

be 3 to 10 times greater than that for spheres of equivalent volume.

4. CONCLUSIONS

Based on the observations made in 3 above, it would appear that our optical remote sensing measurements of S and I_1 and I_r (for scattering angles around 120°) should be severely biased by the presence of irregularly shaped aerosol particles, whereas our multi-wavelength optical depth measurements should be only weakly biased by the presence of such particles. Using spherical particle theory, we might then expect to successfully invert for columnar size distributions from our multi-wavelength optical depth data even if a significant fraction of the aerosol particles were non-spherical. However, it would appear very questionable whether we could successfully retrieve a meaningful particle size distribution or refractive index from our measurements of S and I_1 and I_r for such a condition. In particular, it would appear doubtful whether there would be much similarity in the shapes of the size distributions obtained from the multi-wavelength optical depth and I_1 and I_r angular scattering data. Radiative transfer calculations of spectral diffuse – direct flux ratios made using aerosol parameters inferred from our combined remote sensing techniques would also appear unlikely to agree very well with our spectral flux sensor measurements.

In analyzing our measurement results to date, we have not encountered the above mentioned difficulties. In contrast, we have obtained very acceptable consistency in our various remote sensing inferences of the aerosol size distribution and particle refractive index. Recognizing that atmospheric aerosol particles are not all perfect spheres, we are led to conclude that non-spherical particles do not bias remote sensing determinations of atmospheric aerosol properties as significantly as laboratory measurements of non-spherical particle scattering indicate.

Acknowledgements. The work reported in this paper has been supported by the Atmospheric Sciences Section of the National Science Foundation under Grant ATM 75-15551A01.

Question (CUZZI):   What was the cross section weighted mean radius for your distributions? If $\bar{x}$ is less than several, the assumption of sphericity may not be too bad, especially if the particles are liquid.

Answer (REAGAN):   Results from both concurrent direct measurements and from our size distribution inversions indicate almost no particles with radii greater than 5 µm, and the cross section (extinction) weighted mean radius is typically less than 1 µm. For a midvisible wavelength, this means that $\bar{x}$ is typically less than 10.

Question (SHETTLE):   Getting back to a point raised by Dr. Kerker, most of the data you showed was for refractive indexes appropriate for liquid particles, which you would expect to be reasonably spherical.   On days when there has been a lot of blowing dust and you infer a refractive index on the order of 1.5 or more, do you still have reasonable consistency between your different measurement techniques?

Answer (REAGAN):   On the days where we have been able to collect a reasonably complete set of measurements, we have found good consistency in our results.   Unfortunately, we have yet to collect a complete set of observations under really strong blowing dust conditions.   However, direct particle measurements made during some of our observations indicate that at least 20% of the particles were non-spherical with major to minor axis ratios of 2 or greater.

REFERENCES

Hodkinson, J. R., 1963, *in* "Electromagnetic Scattering," M. Kerker, Ed., Pergamon Press.

Holland, A. C., and Draper, J. S., 1967, *Appl. Opt.*, *6*, 511.

Holland, A. C., and Gagne, G., 1970, *Appl. Opt.*, *9*, 1113.

King, M. D., and Herman, B. M., 1979, *J. Atmos. Sci.*, *36*, 163.

King, M. D., Byrne, D. M., Herman, B. M., and Reagan, J. A., 1978, *J. Atmos. Sci.*, *35*, 2153.

Perry, R. J., Hunt, A. J., and Huffman, D. R., 1978, *Appl. Opt.*, *17*, 2700.

Pinnick, R. G., Carroll, D. E., and Hofmann, D. J., 1976, *Appl. Opt.*, *15*, 384.

Pollack, J. B., and Cuzzi, J. N., 1978, *in* "Preprints of 3rd Conf. on Atmospheric Radiation," Davis, CA, June 1978, Amer. Meteor. Soc.

Reagan, J. A., Byrne, D. M., King, M. D., and Herman, B. M., 1977, *in* "Preprints 4th Joint Conf. on Sensing Environmental Pollutants," New Orleans, Nov. 1977, Amer. Chem. Soc. (submitted for publication in revised form to *J. Geophys. Res.*).

Shaw, G. E., Reagan, J. A., and Herman, B. M., 1973, *J. Appl. Meteor.*, *12*, 374.

Spinhirne, J. D., 1977, Monitoring of Tropospheric Aerosol Optical Properties by Laser Radar, Ph.D. dissertation, Univ. of Ariz., Tucson, Ariz.

Spinhirne, J. D., Reagan, J. A., and Herman, B. M., 1979, Vertical Distribution of Aerosol Extinction Cross Section and Inference of Aerosol Imaginary Index in the Troposphere by Lidar Technique, (submitted for publication in *J. Appl. Meteor.*).

Zerull, R. H., and Giese, R. H., 1974, *in* "Planets, Stars, and Nebulae Studies with Photopolarimetry," T. Gehrels, Ed., Univ. of Ariz. Press.

PARTICIPANTS

Acquista, Charles	Department of Physics, Drexel University, Philadelphia, PA 19104
Aronson, James	Arthur D. Little Company, 15 Acorn Park, Cambridge, MA 02140
Asano, Shoji	Geophysical Institute, Tohoku University, Katahira-2 chome, Sendai, Japan
Ashkin, Arthur	Bell Laboratories, Holmdel, NJ 07733
Barker, Ewa	Institute on the Environment, State University of New York at Albany, Albany, NY 12222
Bickel, William	Department of Physics, University of Arizona, Tucson, AZ 85721
Blanchard, Duncan	Atmospheric Sciences Research Center, State University of New York at Albany, Albany, NY 12222
Bottiger, Jerold	Department of Physics, Texas A & M University, College Station, TX 77843
Bunting, James	Air Force Geophysics Laboratory, Hanscom Air Force Base, MA 01731
Cheng, Roger	Atmospheric Sciences Research Center, State University of New York at Albany, Albany, NY 12222
Chýlek, Petr	Atmospheric Sciences Research Center, State University of New York at Albany, Albany, NY 12222 and National Center for Atmospheric Research, Boulder, CO 80307

Cuzzi, Jeffrey   Space Science Division/MP 245-3, Ames
                               Research Center, Moffett Field, CA 94035

Daum, Gaelen   Ball Research Laboratory, Mail Code DR-
                               DAR-BLB, Aberdeen Proving Ground, MD 21005

Deepak, Adarsh   Institute for Atmospheric Optics and Remote
                               Sensing, P.O. Box P, Hampton, VA 23666

Deirmendjian, Diran   The Rand Corporation, 1700 Main Street,
                               Santa Monica, CA 90406

Detenbeck, Robert   Physics Department, University of Vermont,
                               Burlington, VT 05405

Druger, Stephen   Department of Physics, Clarkson College of
                               Technology, Potsdam, NY 13676

Embury, Janon   U.S. Army Chemical Systems Laboratory,
                               Mail Code DRD-CLB-PS, Aberdeen Proving
                               Ground, MD 21010

Fenn, Robert   Air Force Geophysics Laboratories/OPA,
                               Hanscom Air Force Base, MA 01731

Fry, Edward   Department of Physics, Texas A & M
                               University, College Station, TX 77843

Goedecke, George   Department of Physics, New Mexico State
                               University, Las Cruces, NM 88003

Grams, Gerald   School of Geophysical Sciences, Georgia
                               Institute of Technology, Atlanta, GA 30332

Greenberg, J. Mayo   Huygens Laboratory, University of Leiden,
                               Leiden 2405, Netherlands

Gustavson, Bo   Space Astronomy Laboratory, State University
                               of New York at Albany, Executive Park East,
                               Albany, NY 12203

Herman, Benjamin   Department of Atmospheric Physics,
                               University of Arizona, Tucson, AZ 85721

Hogan, Austin   Atmospheric Sciences Research Center,
                               State University of New York at Albany,
                               Albany, NY 12222

Holland, Alfred   National Aeronautics and Space
                               Administration, Wallops Flight Center,
                               Wallops Island, VA 23337

Holst, Gerald — U.S. Army Chemical Systems Laboratory, Mail Code DRD-CLB-PS, Aberdeen Proving Ground, MD 21010

Huffman, Donald — Department of Physics, University of Arizona, Tucson, AZ 85721

Jayaweera, Kolf — Meteorology, Atmospheric Sciences, National Science Foundation, Washington, DC 20550

Jiusto, James — Atmospheric Sciences Research Center, State University of New York at Albany, Albany, NY 12222

Kattawar, George — Department of Physics, Texas A & M University, College Station, TX 77843

Kerker, Milton — Department of Chemistry, Clarkson College of Technology, Potsdam, NY 13676

Kiehl, Jeffrey — Department of Atmospheric Sciences, Earth Science Building, State University of New York at Albany, Albany, NY 12222

Ko, Malcolm — Atmospheric and Environmental Research, Cambridge, MA 12138

Lauer, James — Department of Mechanical Engineering, Rensselaer Polytechnic Institute, 110 8th Street, Troy, NY 12181

Lewis, Lonzy — Atmospheric Sciences Research Center, State University of New York at Albany, Albany, NY 12222

Liou, Kuo-Nan — Department of Meteorology, University of Utah, Salt Lake City, UT 84112

Loesch, Arthur — Atmospheric Sciences Research Center, State University of New York at Albany, Albany, NY 12222

McKechney, William — Air Force Office of Scientific Research/NC, Bolling Air Force Base, Washington, DC 20332

Misconi, Nebil — Space Astronomy Laboratory, State University of New York at Albany, Executive Park East, Albany, NY 12203

Moroz, Eugene   Air Force Geophysics Laboratory/LYU,
                             Hanscom Air Force Base, MA   01731

Mukai, Sonoyo   Kanazawa Institute of Technology, P.O.
                             Box Kanazawa-South, Ishikawa, 921 Japan

Pearce, William   Washington Analytical Service Center of
                             EG & G, 6801 Kennilworth Avenue, P.O. Box
                             398, Riverdale, MD   20840

Pinnick, Ronald   U.S. Army Atmospheric Sciences Laboratory,
                             White Sands, NM   88002

Ramaswamy, V.   Atmospheric Sciences Research Center,
                             State University of New York at Albany,
                             Albany, NY   12222 and National Center for
                             Atmospheric Research, Boulder, CO   80307

Ratcliff, Keith   Space Astronomy Laboratory, Executive Park
                             East, Albany, NY   12203 and Department of
                             Physics, State University of New York at
                             Albany, Albany, NY   12222

Reagan, John   Department of Atmospheric Physics,
                             University of Arizona, Tucson, AZ   85721

Reilly, Edwin   Department of Computer Science, State
                             University of New York at Albany,
                             Albany, NY   12222

Rogers, Christopher   Department of Astronomy, University of
                             Toronto, Toronto, Canada   M5S1A7

Sato, Makoto   National Aeronautics and Space
                             Administration, Goddard Space Flight Center,
                             Institute for Space Studies, 2880 Broadway,
                             New York, NY   10025

Saunders, M.J.   Bell Laboratories, 2000 Northeast
                             Expressway, Norcross, GA   30071

Schaefer, Richard   Space Astronomy Laboratory, State University
                             of New York at Albany, Executive Park East,
                             Albany, NY   12203

Schuerman, Donald   Space Astronomy Laboratory, State University
                             of New York at Albany, Executive Park East
                             Albany, NY   12203

Shettle, Eric	Air Force Geophysics Laboratory, Atmospheric Optics Branch, Hanscom Air Force Base, MA 01731
Shirkey, Richard	U.S. Army Atmospheric Sciences Laboratory, White Sands, NM 88002
Srivastava, Madhu	Department of Electrical Science, State University of New York at Stony Brook, Stony Brook, NY 11794
Stein, David	Box 1371, Branch P.O., Rome, NY 13440
Stuebing, Edward	U.S. Army Chemical Systems Laboratory, Mail Code DRDAR-CLB-PS, Aberdeen Proving Ground, MD 21010
Toller, Gary	Space Astronomy Laboratory, State University of New York at Albany, Executive Park East, Albany, NY 12203
Trusty, Gary	Naval Research Laboratory, Code 5568, Washington, DC 20375
Turner, Robert	Science Applications, Inc., 15 Research Drive, Ann Arbor, MI 48103
Vonnegut, Bernard	Atmospheric Sciences Research Center, State University of New York at Albany, Albany, NY 12222
Wang, Dausing	Department of Physics, Clarkson College of Technology, Potsdam, NY 13676
Wang, Ru	Space Astronomy Laboratory, State University of New York at Albany, Executive Park East, Albany, NY 12203
Weil, Herschel	Department of Electronic and Computer Engineering, University of Michigan, Ann Arbor, MI 48109
Weinberg, Jerry	Space Astronomy Laboratory, State University of New York at Albany, Executive Park East, Albany, NY 12203
Weiss, Karin	Ruhr Universität Bochum, Bereich Extraterrestrische Physik, Gebaude NA-01, 4630 Bochum, West Germany

Welch, Ronald

Department of Atmospheric Sciences,
Colorado State University
Fort Collins, CO 80523

Wiscombe, Warren

National Center for Atmospheric Research,
P.O. Box 3000, Boulder, CO 80307

Woods, David

National Aeronautics and Space
Administration, Langley Research Center,
Mail Stop 475, Hampton, VA 23665

Worden, Robert

Logica Ltd., 64 Newmans Street,
London W1 4FE, United Kingdom

Yeh, Cavour

Department of Engineering, University of
California at Los Angeles, 405 Hilgard
Avenue, Los Angeles, California 90024

Zerull, Reiner

Bereich Extraterrestrische Physik, Ruhr-
Universität, Bochum, West Germany

SUBJECT INDEX